T5-AEY-616

SAUNDERS GOLDEN SERIES IN ENVIRONMENTAL STUDIES

ECOLOGY, POLLUTION, ENVIRONMENT by Amos Turk, Jonathan Turk, and Janet T. Wittes

ENVIRONMENTAL SCIENCE by Amos Turk, Jonathan Turk, Janet T. Wittes and Robert Wittes

ECOSYSTEMS, ENERGY, POPULATION by Jonathan Turk, Janet Wittes, Robert Wittes and Amos Turk

SCIENCE, TECHNOLOGY and the ENVIRONMENT by John T. Hardy

IMPACT: SCIENCE ON SOCIETY edited by Robert L. Wolke

SCIENCE, MAN AND SOCIETY *(Second Edition)* by Robert B. Fischer

OUR GEOLOGICAL ENVIRONMENT by Joel S. Watkins, Michael L. Bottino and Marie Morisawa

This book has been printed on 100% recycled paper

IENCE ON SOCIETY IMPACT: SCIENCE ON SOCIETY IM

CT: SCIENCE ON SOCIETY IMPACT: SCIENCE ON SOCIETY

SAUNDERS GOLDEN SERIES

Impact:

ENCE ON SOCIETY IMPACT: SCIENCE ON SOCIETY IMPA

SCIENCE

: SCIENCE ON SOCIETY

ON

IMPACT: SCIENCE ON SOC

SOCIETY

1975 W. B. SAUNDERS COMPANY / PHILADELPHIA / LONDON / TORONTO

W. B. Saunders Company: West Washington Square
Philadelphia, PA 19105

12 Dyott Street
London, WC1A 1DB

833 Oxford Street
Toronto, Ontario M8Z 5T9, Canada

Cover photograph is "Science and Mankind," enamel on steel mural by Virgil Cantini, University of Pittsburgh

Library of Congress Cataloging in Publication Data

Main entry under title:

Impact, science on society.

(Saunders golden series)

Includes index.

1. Science – Social aspects. I. Wolke, Robert L.

Q175.5.I56 301.31 74-12920

ISBN 0-7216-9585-X

Impact: Science on Society ISBN 0-7216-9585-X

Last digit is the print number: 9 8 7 6 5 4 3 2 1

PREFACE

This is a book about science, but it is not a science book in the usual sense. And while it is intended primarily for use in college-level non-science major courses in chemistry, physics, biology, environmental sciences, earth sciences or general science, it is not a textbook on any of these subjects. It is a book to be used *in conjunction with* the textbooks on these sciences, to fulfill the vital function of showing the student how that science and its resulting technology are interacting with and changing—for better or worse—the society in which he lives. It focuses, not on the science itself, but on the societal impact of science, and it presumes no scientific or technical sophistication on the part of the reader.

In almost every college or university science course there is a textbook that presents the subject-matter of the course: the facts and theories and principles that are to be learned by the student. The student studies the textbook, and he attends lectures, multimedia presentations, and recitation and laboratory sessions, and he may take examinations and write reports, after which he emerges with a grade that, at least in principle, reflects his degree of mastery of "the material." And that, usually, is that. And onward to the next course.

But especially if the course in question is a science course, and in particular if it is one of the very limited number of science courses that a student will ever experience in college because of the fact that he is not majoring in science, this traditional sequence of operations is not enough. It is not enough that a French major master some principles of chemistry, or that a sociology major know some biology, or that a music major perform some physics experiments in an attempt to satisfy the "general education" or "distribution of studies" requirement, which is imposed by colleges and universities for the purpose of ensuring a minimal degree of breadth to their students' educations. All too often, when administered in the usual way, these courses in science for nonscientists provide not breadth, but only a selection of other narrownesses to supplement the narrowness of the student's own major. The expectation seems to be that the student himself will fill in the spaces between these narrow peaks of mastery, that he himself will ultimately realize the relationships that have grown up in our late twentieth-century society between biology and morality, for example, or between physics and politics, or between chemistry and the survival of the world's still-starving peoples; and that he will somehow emerge a liberally educated, sensitive and perceptive citizen. It is, however, a cold and unfortunate fact that the vast majority of students will never fill in these spaces for themselves; they will never perceive these critical interrelationships—not because of laziness or inability, but just because the atmosphere and incentive for the necessary contemplation are not provided by a curriculum that consists only of one "subject-matter" course after another. True liberalism in education cannot come from a series of courses on disconnected subjects. The interrelationships among them must be taught as explicitly as the subjects themselves.

Science as a discipline is unique in this respect. Other disciplines that are included in college and university curricula are more or less passive; they are studies of various aspects of human society—its arts, its philosophies, its history, and so on. But science is an *active* discipline—science and its miraculous technology are actively and constantly *changing* our society. To teach science only as a body of knowledge or even as a facet of culture, therefore, is to ignore a vital educational function: that of showing the student, especially the student not majoring in science, how the science he is studying relates to and affects the greater operation of the society in which he lives. This interrelationship of science and society must be a deliberate part of at least one of the college or university science courses taken by the nonscience major. Doing this goes far beyond the mere inclusion of relevant and currently popular topics among the chemical, biological or physical themes that are covered in a science course. It even goes beyond the substitution of whole new courses for the traditional ones, such as environmental science for traditional chemistry or ecology for traditional biology. These are only new and different, albeit more appropriate, "subject matters." What must be learned by the nonscientist during his brief, few-credits encounter with the sciences in college is how that science—and indeed how *all* of science and technology—is affecting the physical, mental and spiritual human condition in our twentieth-century society, and what promises—and terrors—it may hold for the future.

How can this urgent educational job be done by the harassed college or university professor who wishes he had enough time to do his research and subject-matter teaching better than he does, much less to read omnivorously on all the far-reaching implications his science may hold for world society, he might present them in balanced perspective to his students? The impact of science on society has been and continues to be so far-reaching that it is becoming a discipline in itself: there is at least one large university program in the study of Science, Technology and Society; and the new field of Technology Assessment is attracting the attention of a growing number of people in the academic world, in federal agencies, and in Congress. The college or university science teacher simply hasn't the time to dig out, to organize and to assess the burgeoning literature on these vital societal interactions, which were either completely missing from or only peripheral to his own education when *he* was in school.

This book is intended to enable the instructor to do that job: to liberalize and generalize the science education he is providing for his students at little burden to his own resources and at little sacrifice to the technical subject matter that both he and his curriculum planners feel is minimally essential for the liberal arts student. For that student it is hoped that this book, used as suggested below, will stimulate the necessary cross-currents of thought that will tie the technical subject matter down in his mind to what he feels are the "real" human issues—the issues he sees on television and in the movies and reads about in the newspapers.

This book doesn't teach the science itself: the chemistry or physics or biology. It is presumed that that function is already being accomplished in the classroom and laboratory through the inclusion of any collection of scientific topics from the traditional to the highly relevant, and by utilizing any teaching methods from the traditional to the highly innovative. This book is to be used in *parallel* with those activities. It has been written in the most general and nontechnical way possible, so as not to wed it to any specific syllabus, although in rare places, when a bit of technical explanation was deemed necessary to the understanding of the article, this has been added editorially.

Before suggesting practical ways in which this book can be used to accomplish its stated purpose, a few words should be said about the origin of both the original and reprinted material that it contains. Important information concerning the impact of science on society lurks in sources as diverse as popular magazines, foundation reports, congressional testimonies and graduation addresses. For one person to produce a science-and-society reader by collecting a representative sampling of thought from the spectrum of literature in this vast field would be both prohibitively difficult and presumptuous. Moreover, such a collection would be useless to the student without someone's organizing it for him and putting it all into meaningful perspective—again a presumptuous effort for a single author or editor. Hence, the method that was adopted was to enlist the abilities of some highly knowledgeable people, each in a specific area of the science-society interface, and to ask them to select those articles by other experts that best characterize the most important interactions within their areas. The personal insights of these knowledgeable people—our Contributing Editors—were not wasted by limiting their rôle to bibliographer, however, since each one was asked to set the scene by writing an original introduction to his section (titled "The Problem") and a summary of his own impressions of the prospects for the future (titled "Prognosis"). Thus, the book draws both upon the best available writings by active experts in the science-technology-society arena and upon the perspective evaluations of scholarly observers who, it must be confessed, are active experts in their own right.

And now, how can this book be used? There are three general ways.

1. Primarily it is intended to be used in nonmajor science courses as a supplementary reader. Although the mechanism used will depend to a great extent on the individual instructor, perhaps the most direct is to assign selections for at-home reading and then, for an hour every other week, to conduct an informal discussion on the assigned material. It is, after all, the *discussion* that is important. The material in this book has been designed insofar as possible to be down-to-earth, stimulating, thought-provoking and controversial; students should *react* to it rather than accept it as dogma. It is in these discussion sessions, whether held

during scheduled recitation time or in someone's living room or under a tree, that the students will feel free to argue the villainy or virtue of science in any particular context, and in so doing will form the associations and viewpoints which *to them personally* represent the rôle of science in society. Toward this end, the students should be encouraged to mark with highlighter pens those passages in the book that particularly turn them on or off, to be used as sparks for igniting later discussion. And, without any intended offense, may it be suggested that the instructor himself can learn from both the readings and the ensuing discussions, for it is this kind of informal but substantive "rapping" with one's students that can be one of the most stimulating and satisfying rewards of teaching.

In making the reading assignments, incidentally, the instructor will find that the book progresses from highly practical matters in the earlier sections to more philosophical matters in the later ones. The sections are completely independent of one another, however, and can be assigned in any order that seems best to complement the course's in-class subject matter.

2. This book can be used as the primary source for an actual course on those contemporary societal problems that stem from the burgeoning development of science and technology. Such a course could be structured either around the technical roots of the problems, in which case it would be a "contemporary chemistry" (or physics or biology) course, or around the ways in which society is trying to cope with them, in which case it would be a "contemporary problems" course. In the former case, this book would serve as a thread upon which the technical details are to be strung; in the latter, it would be the textbook (or one of the textbooks) itself.

3. In the history of literature, no author has been able to repress the conviction that his book should be read by almost everyone. It would be presumptuous of the present author to claim to be an exception. His conviction has been strengthened, however, by the reactions of many nonscientist friends and acquaintances who, upon hearing what this book was to be about and upon being assured that it wasn't to be a "science book" full of equations and formulas and other incomprehensibles, have expressed an interest in reading it to satisfy a long-felt desire to know "what science is all about" without having to plow through the forbidding stuff itself. Thus, it is hoped that almost any generally aware, nontechnically oriented citizen of our oft-bewildering technological society—college educated or not—will find in this book some of the insight he or she needs to evaluate the unending stream of news concerning scientific and technological developments and their associated problems.

Finally, I wish to express my gratitude to some of the people who helped to make this book possible: first, of course, to the Contributing Editors, who added judgment, authority and dimension to their delegated topics; to Tony Griffiths, who gave me the idea for this book; to the staff of the W. B. Saunders Company, particularly to Joan Garbutt for her encouragement and to Lorraine Battista for some excellent and sensitive design work; to Virgil Cantini for his gracious cooperation in letting us adapt his mural for use as a cover; to Herb Ferguson for his generosity and skill in photographing the mural; and to my secretary, Patricia Moan, who typed parts of the manuscript, tracked down essential information and handled the exasperating job of coordinating the copyright permissions for all the reprinted material.

Robert L. Wolke
July, 1974

CONTENTS

IMPACT: SCIENCE ON SOCIETY

PROLOGUE

In case you wonder why it is
We're reaching toward the stars
With satellites and Skylab flights
And probes to Moon and Mars,
While spending all those billions for
The depths of space to plumb,
Just look at our environment
And see what Earth's become.

We've fouled up all our rivers and
Our lakes and brooks and streams.
What once was clear now looks like beer,
Except more suds, it seems.
The Rhine is radioactive and
The Blue Danube is green,
And swimming in Lake Erie will
Remove you from the scene.

The rain's sulfuric acid and
The snow, it looks like mud.
The Mississippi River banks
Are all choked up with crud.
The stuff you see from on the quay,
It certainly is weird.
I've walked upon the waters, and
Don't even have a beard!

The acid waste and sewers flow
In widening rivulets.
The farther down the stream you go,
The filthier it gets.
The water that we drink depends
On what's upstream, and who.
The kids take water from the tap
To sniff instead of glue!

With DDT and mercury
And phosphates all around,
Our friends the fish have just one wish:
To keep from being drowned,
Asphyxiated, strangulated,
Lest to air they go.
(But now I come to think of it,
They're better off below!)

The fact'ries belch out smoke and soot
Until our lungs are choked.
I'm getting emphysema, and
I've never even smoked!
The air is full of SO_2
And H_2S to boot.
I'd kill the damned polluters, but
I can't see where to shoot!

The local power plant puts so
Much smoke into the air,
We have to turn on all the lights
To see from here to there,
Which uses so much power that
They generate some more,
Which makes so much more smoke and soot,
It's darker than before!

So, off to Space! Let's go and see
What other worlds may hold.
Not just for curiosity,
Or just because we're bold.
It's exploration that's our game!
So don the old space-suit,
Count down! Lift off! Perhaps we'll find
More places to pollute!

R. L. WOLKE

WORLD IS WARNED
ON RESOURCE USE

Famine Growing

World Faces
Stark Fact:
d Dwindlin

Sonar's

Crisis

Bauxite

VINYL CHLORIDE
PLANTS, EPA

Veterans Still

Satellites May
Make Earth
The Smallest
Of Worlds

COURT TO PERM
POLLUTION BY MINE
IN LAKE SUPERIOR

Brazil
Iceberg

ANNOUNCE
MIC TESTS

FOUR NEW CENTERS
FOR CANCER SET UP

Occurrences
omic Plants,

India Gets Sa
For Reading

ENERGY FROM SUN
SEEN IN 5 YEAR

A. Limits Antacid Contents
Safe and Proven Ingredients

olar Heating Sys

1980s Predicte

4 PA
ARE

Radiation F
In '54

Scientists Debate
Mari
Effects

Mysterious Cancer That Just Stops

G SYSTEM

H-Test
g Toll

r Found
rt Factor

Canadians Battle Ice to Plan
A Gas Pipeli rom Arctic

2 PH.D.'S IN TEXAS
SEEK HOUSE SE

SOFT WAT
NO HEART A

Copper in Pennies
May Be Reduced

Who's Next in A m C

SCIENTIS
TO AID C

Geothermal Power
Hunted in Montana

China Sets Off
Nuclear Blast
In Atmosphe

ss recycles used auto

s Meet.
aft Pact
a Use

4 Power Plants
Called Polluters

Lead Poisoning
Topic of Seminar

A-Bo
The
Fea

n Fact For World:
l Stocks Dwindle

Animal Frozen
Revive

County Pane
Effec
Air P

Food Additive
Danger Claim
Test

New Facts
Artificial D

HOSPITAL PL
FOR CRITIC

aded
Called

Dow Chemical Expects
rter Results

INTRODUCTION

Robert L. Wolke, General Editor

The headlines on the facing page appeared during a two-week period selected at random in the summer of 1974. Similar headlines are rolling off the presses as you read this, and they will continue to do so with increasing frequency and increasing urgency. They differ from other headlines in that they deal not with mere events, but with *problems*—problems that were born as discoveries in science, then nurtured and intensified by the ingenious practical uses to which technology and industry have applied these discoveries, and turned loose upon a society that does not yet know how to cope with them.

For thousands of years, mankind's way of life had remained pretty much the same: a difficult struggle for survival. Here and there individuals, societies and civilizations died for lack of the essentials of food, water or security. And then explosively in the twentieth century, and especially in its latter half, we learned how to do an astounding variety of brand new things, from survival-insuring things like erasing insects, improving our foods and eliminating diseases to more frivolous things, such as flying through the air faster than thunder and travelling to the moon. It appeared that there was no limit to the miracles that we could work, and the world's people, on the average, began to live longer and better lives.

But then sometimes we found ourselves doing miraculous new things for little apparent reason other than that we knew how to do them. And sometimes, even though our objectives were clearly humanitarian and aimed directly at the solution of an urgent human problem, we were dismayed to find that the technological solution to one problem could create other problems: unforseen, undreamed of, and perhaps even worse than the one that had been solved.

And that is where today's headlines come in. Individually, they range from the merely perplexing to the downright frightening. But taken all together, their theme is this: There remains not a single cranny of our everyday lives that has not been permeated by twentieth-century science and technology. We seem to have reached a kind of saturation point, not only in our ability to improve our own condition as human beings, but also in our ability to change Nature itself, so that any further "improvements" we make in our natural surroundings seem to turn suddenly around and threaten us. We appear to be very close to The Limit: the breaking point beyond which our scientific intellect and technological ingenuity cannot push a usually resilient and compliant Mother Nature without the most serious of consequences. A few of these unexpected consequences have recently become painfully obvious, pollution being the prime example.

And while it is now easy to recognize this problem and indeed even to satirize it, as in the cynical verses in the Prologue of this book, we know that it is far easier to satirize than to solve. But our awareness of pollution, for all its complex and tangled roots in technology, politics and economics, is only the first symptom of a growing awareness that has deep moral and ethical overtones: that, to borrow a cliché, "WE CAN'T GO ON LIKE THIS." We have covered over with concrete or asphalt almost one out of every two hundred square feet of the land surface of the United States of America. To make useful things for ourselves, we have taken billions of tons of minerals from the crust of the earth where they had lain undisturbed for three billion years; and we are now beginning to run short of almost everything. We have learned enough about living organisms—including ourselves—that we can almost tailor the basic genetic materials of life into new forms, a process for which we have invented the innocent-sounding name of "genetic engineering." And if our slow demise does not follow from any of these dangerous manipulations, we have the means already at hand to end the human race with a catastrophic thermonuclear bang.

What can individual citizens do besides worry and hope? The foremost necessity is that we become *knowledgeable* and *aware:* knowledgeable about the technical roots of the scientific and technological advances that threaten us, and aware of the problems that our society faces as a result of these advances. For only when armed with both knowledge and awareness can a citizen hope to voice opinions that will be listened to by those who are in a position to act: those in government, in industry and in science.

While the necessary factual and technical knowledge can be supplied in principle by reading science books and by taking science courses, the necessary awareness must generally be developed elsewhere: by a book, it is hoped, such as this. (The purpose and methods of using this book are described in the Preface.)

First, it is important to realize fully the distinction between *science* and *technology.* Science is purely an effort to understand the workings of Nature. Whether carried out in laboratories, in the field or in libraries, the sole objective of science is to obtain knowledge. Technology, on the other hand, is man's *application* of his knowledge to produce useful goods or useful work; it is carried out both privately by industry in factories or mines or operating rooms, and publicly by governments in mission-oriented projects ranging from sewage disposal to nuclear weapons development. Without science our technology could not function, and without technology our science could not improve our lives—or threaten them.

Science's impact on society, then, is exerted mainly through the medium of technology, affecting our lives in a thousand different ways from the moment we are born to the moment we die. (And as we shall see, it has even begun to determine *whether* we are born and *whether* we die.) In the following pages we shall examine the broad spectrum of interactions between science and society—those interactions that have spawned the major problems of our technological age. In doing so we shall concentrate on the problems themselves—not on the technical details of the science at their roots. For the vital importance of science to our lives is not to be communicated to a reader by dazzling him with an endless recitation of scientific and technological miracles; he is already more than willing to concede to science its wizardry of drugs and plastics and computers and satellites. It is, rather, through direct confrontation with the societal problems themselves that the reader can best appreciate the personal importance of science—its impact on everything from his hamburger bun (see page 212) to his grandfather's funeral (page 190).

This book has been divided into five sections, each one covering a major area of concern in the interaction of science with society. **Section One** deals with mankind's supply of matter and energy. How much do we have? How fast are we using it up? Will we run completely out of sources of energy and of the hundreds of mineral substances from which we make everything we need? And what about that first inkling we are beginning to have of the most frightening of all shortages: that the world may be running out of food? For in the final analysis, all we humans have to work with is energy and matter: the former from the sun (mostly via photosynthesis, which made those prehistoric plants that later became our coal and oil) and the latter in the form of whatever we can manage to find and dig out of the earth's crust. Mankind has inherited only the earth. Is he squandering this one-time-only inheritance?

The Contributing Editors who have selected the readings in Section One and who have put them into perspective by writing an introduction ("The Problem") and a summary ("Prognosis") to the Section, are well qualified in both of the necessary fields: energy and minerals. Drs. Irving Wender and Bernard Blaustein are scientists at the Pittsburgh Energy Research Center of the U. S. Bureau of Mines, Department of the Interior. Dr. Wender is Research Director of that Center. His research work has been concerned with catalysis, with the chemistry of coal, and with the conversion of coal into other forms of fuel. He has received high awards from the U. S. Department of the Interior for his work, and is now deeply involved in trying to solve our energy problems. Dr. Blaustein, trained both in chemical engineering and in chemistry, has been interested in teaching as well as in coal research; he has held teaching appointments at the Johns Hopkins University and at the Universities of Maryland and Pittsburgh.

Section Two is concerned with the vast outpouring of "stuff and things" produced by our industrialized society: the astounding variety and number of products we have learned to make—and to want more of. Is our technological know-how a bountiful cornucopia, or is it burying us in junk? In the United States we are producing billions of tons of goods per year, much of it destined only to be used up, thrown away and replaced. But besides the mountainous disposal problem generated by all these materials, there exist some of those totally unexpected problems that we alluded to earlier: the manufacturing activities involved in producing all these goods are not only polluting our air and water, but can also inject insidious and sometimes unsuspected poisons into our lives: for example, the mercury and the lead we have all read about and possibly the "PCBs" we have not read much about—yet.

The Contributing Editor who chose the selections for Section Two, Dr. A.

Truman Schwartz, is a physical chemist, a former Rhodes Scholar with a master's degree from Oxford University and a doctorate from M.I.T. He is the author of a recent college textbook in chemistry for nonscientists, which demonstrates his flair for explaining complex technical problems to nontechnically-trained people. As his "Problem" and "Prognosis" essays will show, Dr. Schwartz remains optimistic about man's ultimate ability to manage his technological powers.

While the sheer magnitude of some of man's activities threatens to overpower our environment's ability to absorb and adjust to them, it is not only the *scale* of the operations that can cause problems. As is shown in **Section Three,** there are sometimes unexpectedly large natural consequences of what might appear to be an activity of only modest size. One example is the spraying of crops with insecticides, which has been found to cause unexpected shifts in the delicate balance of Nature. And it is only our slowly dawning realization that certain small impositions on Nature can trigger dangerously large—even global—effects, that has kept the United States from dashing headlong into the construction of a fleet of supersonic commercial airliners (the so-called "supersonic transports," or "SSTs"). Who would have dreamed, even a few decades ago, that an airplane could be accused of causing skin cancer (page 136)?

The Contributing Editor of this Section on how man's activities are modifying Nature for better or for worse is Scientific Director of *Environment* magazine.

In **Section Four** we turn to more personal concerns, concerns that strike at our very being as humans—at our bodies, our minds, our ethics and our morals, at the quality of our lives and of our deaths. For science and technology—medical, biological and psychological technology in this case—are busily changing and manipulating these things too, again for better or for worse. The science fiction of a generation ago, in which babies are created and grown in laboratory vessels, human thoughts are controlled through electrodes implanted in people's brains and genetic materials are manipulated to make a super-race of men, is on the verge of becoming a reality. Already we are wrestling with the ethical and moral dilemmas brought on by our impending ability to act as gods. Even as this book goes to press, scientists have called a voluntary halt to certain genetic manipulation experiments in which new types of living bacteria and viruses could be created—types which have never existed before in the universe and which, if not scrupulously controlled, might be able to infect and destroy an utterly defenseless mankind. As one of the scientists involved said, "This is the first time I know of that anyone has had to stop and think about an experiment in terms of its social impact and potential hazard."

In Section Four, then, we stop to ponder the ethics and morality of the modern life sciences. The Contributing Editors, Drs. Rustum Roy and G. R. Barsch, an Indian-born chemist and a German-born physicist, respectively, are both at the Pennsylvania State University's Materials Research Laboratory, of which Dr. Roy is Director. In addition to his extensive scientific research on the structures and properties of solid materials, Dr. Roy has lectured and written widely on the interactions of religion and philosophy with contemporary society. Having also served on many advisory committees for the State of Pennsylvania, the National Academy of Sciences, the National Science Foundation and the National Council of Churches, he is particularly well qualified to comment on the ethics and morality of modern science.

Finally, in **Section Five** we examine the broad prospects for science as a human activity. Does science have any function beyond the spawning of useful technologies? Are scientists "different" from other people? Do they arrogantly disregard the social consequences of their work? What has science done for society in the past? What can or can it not do, and more importantly, what *must* or *must not* it do in the future? These questions are perhaps even harder to answer than some of those dealt with in earlier Sections. To answer them requires the combined thinking of philosophers, of statesmen and of scientists themselves. Dr. Laurence E. Strong, the Contributing Editor of the final Section of this book and a scientist who has for many years been personally involved in the interaction of science with governments, both in the United States and abroad, has assembled an inspiring collection of such thoughts, as expressed both by observers of science and by leaders and statesmen of the scientific world. And included in such considerations must inevitably be the ultimate and most chilling question of all: How real and imminent is the possibility of a catastrophic end to the human race?

The impact of science on twentieth-century society has become terribly complex. While we have tried to separate the problems into five broad areas, they are really inseparable; references to certain problems inevitably crop up during discussions of others, and the discussions in this book are no exception. But to the student of science, to the student of society, to the ordinary reader of newspapers, it is hoped that the carefully selected readings in this book, together with the personal evaluations of the Contributing Editors, will provide an improved understanding of what science is doing in the world and of its central role in determining the fate of mankind.

Impact:
Science on Society

Photo courtesy of NASA.

SECTION ONE

INHERIT THE EARTH

(USDA photo)

The Problem

Awakened by the annoying brrrrr of the alarm clock (steel springs and genuine plastic cover, powered by electricity carried 19 miles to the house over copper wires). *"Damn clock!,"* I said to myself. Out of bed slowly, from underneath the blanket (also electric), put on my bathrobe (65 percent polyester, 35 percent cotton) and leather slippers. Down the nylon-carpeted stairs and across the vinyl-asbestos-tiled kitchen floor to let in Cat, whom I find stretching to get the cold out of his muscles. *Morning, Cat. Let's eat!* Move the thermostat's servocontrol up a few notches. Hear the furnace (150,000 British thermal units per hour), sitting on the cold basement floor (lime-silica-clay-slag cement) kick on. *Glad those refineries made a lot of heating oil this winter.*

Open the freezer (electric, of course) to get a can of orange juice, and empty it into the pitcher (soda-lime-silica-glass). Add 1 . . . 2 . . . 3 cans of water. *Really too much chlorine in the water these days. Doesn't help the taste any.* Throw aluminum pull-top tab into the bag for the recycling center. Cut out the bottom of the can, (tin-plated steel), throw that in the bag too. Plastic-coated paper body of can goes into the trash. *That's the way! Recycle the metal, burn the rest. Wish there were more recycling nuts like me!*

Pour dry cat food into Cat's dish. Glance at label: "Ground wheat and corn, soybean, corn gluten, wheat germ and fish meals, plus vitamin and mineral supplements. 30 percent protein. Guaranteed to keep his fur glossy". *Anything I can't stand, it's a dull cat!* For me, take out the margarine ("partially hydrogenated vegetable oil, high in polyunsaturated fat, supplemented with vitamins A and D"), one egg, and nonfat milk. Whole wheat bread into the toaster, two slices, with calcium propionate added to retard spoilage. *Thank you, Lord, for our daily calcium propionate.* Put the coffee into the stainless steel (18 percent chromium, 8 percent nickel) coffee pot, plug it in (more electricity), turn on the gas underneath the griddle (teflon-coated carbon steel). Egg sizzles slowly, sunny side up. *Hmm. Reminds me of back when we could see the sun most of the time.*

Walk to the front door to get the newspaper. *Should start thinking about Wednesday's lecture. How can I explain the importance of energy and minerals to my class? A lot of dry facts and numbers? No. Have to think of things that are really meaningful to my college students. Gotta make them see how completely we depend on energy and minerals in everything we do.* Headline: ASSOCIATED PRESS REPORTS SMALL CARS USE 37 PERCENT LESS GASOLINE THAN BIG ONES. PETROLEUM IMPORTS REDUCED. Headline: NATIONAL AIR POLLUTION DROPS SLIGHTLY, EPA SAYS. Headline: NEW GASIFICATION PLANT IN WYOMING TO CONVERT COAL INTO CLEANER-BURNING "SUBSTITUTE NATURAL GAS." Headline: COPPER CARTEL MEETS IN CHILE. PERHAPS TO FOLLOW LEAD OF OPEC. Headline: MORE NUCLEAR ELECTRICITY ON WAY. BUT TOTAL ELECTRIC DEMAND UP ONLY 3 PERCENT OVER LAST YEAR. *Maybe I'll throw in a few statistics after all. They sure speak for themselves. Plus maybe an opinion or two to show them where I stand.*

WEDNESDAY, 9 A.M., LECTURE HALL 12A

Good Morning, class. Today we begin our discussion of energy and minerals. Another name for this topic could be "The State of Our Resources."

You and I have always considered the United States to be a rich country, with all the resources that we need. And, in general, this has been true. Our natural wealth has been one of the main factors responsible for our very high standard of living—still one of the highest in the world in spite of our inflation problems. Until recently, everyone thought we had all the energy and minerals we could possibly use. We've become used to having enough of everything. But, in addition to recent talk about an "Energy Crisis," there is now also the real possibility of a "Minerals Crisis." Let me give you some figures to show what's been happening.

In 1973 we used about 570 million tons of coal in this country, about 6.3 billion barrels of petroleum products, and almost 23 trillion cubic feet of natural gas. In addition, we used sizable amounts of electricity, generated in both hydroelectric and nuclear power plants. All together, this added up to about 75 quadrillion—75,000,000,000,000,000 or 7.5 $X10^{16}$—British thermal units (BTU) of energy. This is certainly a lot of energy, which we used to do an enormous amount of work because, of course, work is what energy is used for—jobs we can't or won't use muscle power for. If we had tried to use muscle power to do all this work instead of using energy from coal, oil, gas, water, nuclear fission and so on, each one of us would have needed at least a hundred slaves—and having slaves, as you may have heard, is out of style. *Well, at least a couple of them snickered at that one.*

Coal, petroleum and natural gas are what we call "fossil fuels", which were formed from the remains of plants and animals that lived

hundreds of millions of years ago. The numbers which are needed to express our consumption of fossil fuels are so huge that they're almost unreal and wouldn't mean a thing to you, so I've done a little calculation with them to put them on a personal basis. If we add up all the tons of coal, barrels of petroleum and cubic feet of gas we used in 1973 and divide by our population, each and every person in this country used or had used for his benefit more than 9 tons of fossil fuel! This figure, 18,000 pounds, is what is called our annual per capita fuel consumption. Our total national energy consumption of 75 quadrillion BTU would have been enough for eighteen million round trips to the moon in an automobile, if such trips were possible. *Now watch that eager beaver in the front row ask me how many miles per gallon a car gets in space!*

"Consumption," of course, is just a fancy way of saying that we burn this fuel up. We burn it to keep our homes and stores warm, and to cook with. We burn it to make steam to make electricity, and to run the factories which made the thousands of things we need and use. And we burn it as fuel in our more-than-a-hundred-million cars, trucks, trains, and airplanes. And some coal and petroleum ends up as plastics and other so-called petrochemical products, and as the tar and asphalt on which we drive.

Besides coal, oil and gas, we use a lot of inorganic minerals in this country, too—materials we dig out of the ground for one purpose or another. Most of this stuff is building material—sand, gravel, stone, cement, and clay—but we're also talking about all of our iron and steel and our aluminum, copper, nickel, zinc, lead, silver, and all the other metals that we need for all the things that keep our American Way of Life going. Adding all these together, in 1973 each person in this country used or had used for him about eleven tons, 22,000 pounds, of minerals. This total amount of material is the equivalent of building 400 Great Pyramids. In 1973 alone!

Nine tons of organic fuels plus eleven tons of inorganic minerals—twenty tons per year—is a lot of stuff for one person, even an American, to use. And this consumption may double by the year 2000! Our country has only five and one-half percent of the world's population, but we use about a third of the whole world's production of fuels and minerals. This is one reason that our standard of living is so much higher than most of the rest of the world.

The United States, of course, is an industrialized or "developed" country. Even overdeveloped in some respects. This means that we use a lot more fuel and mineral resources than does a person in an underdeveloped country. But it also means that even with all our own natural resources, we have to depend on other countries to supply us with some of what we need yet don't produce ourselves. And many of the countries from which we import the petroleum and minerals we use are underdeveloped countries who themselves want to "develop" as we have done. The "have-nots" in the world are very eager to become as wasteful as we are!

Because of The Pill and other "tricks-of-the-trade", our rate of population growth has slowed in the last few years, though our population is

still increasing. And the rest of the world—where the population is increasing more rapidly—will understandably insist on raising their standard of living even faster than we've been raising ours, in order to catch up. It's clear, then, that we're going to have to be much more aware of how dependent we are on one another, because we—the world's people—are already using up our resources of fuels and minerals at an astonishing rate and seem to be trying very hard to use them up even faster.

In addition to our concern about using up our supply of fuels and minerals, there is another aspect to our use of these resources—environmental pollution. Because every bit of mining and refining, and every bit of the living that we do affects the environment. *That sure perked up some ears. I finally said the magic words "pollution" and "environment."*

Let's look at just one part of this enormous problem—air pollution caused by burning fossil fuels. We certainly need the energy from these fuels but, you might say, "There ain't no such thing as a clean and easy BTU." It is a hard fact that the combustion of fossil fuels is our main source of air pollution. There are four major air pollutants that come from the burning of fuels: carbon monoxide, CO; oxides of sulfur, (SO_x); oxides of nitrogen, (NO_x); and particulate matter, which is a fancy name for ash and soot. Note that in these formulas *x* stands for some small number: SO_2, of course, is sulfur dioxide, and SO_3 is sulfur trioxide. Similarly, there are several nitrogen-oxygen gases with various values of *x* in their formulas.

Natural gas is the cleanest, least-polluting fuel that we have. It is practically pure methane, CH_4. When this is burned to provide heat, the reaction can be said in words as: each molecule of methane reacts with two oxygen molecules to yield the combustion products carbon dioxide (one molecule) and water (two molecules) plus heat. Or, in symbols,

$$CH_4 + 2\,O_2 \rightarrow CO_2 + 2\,H_2O + \text{heat}$$

Look at them all writing down the equation. First time some of them have touched pencil to paper since the lecture began! **Natural gas, then, is a clean fuel because neither CO_2 nor H_2O is considered to be a pollutant; you're breathing both of these out into the atmosphere yourself, right now.**

But coal is another fuel that we use in large amounts, and it's another story. Coal is a sedimentary rock made up of carbon compounds, including many different kinds of organic chemicals containing benzene rings (rings of six carbon atoms joined together) and partially hydrogenated benzene rings (the same, with some extra hydrogen atoms added) plus varying amounts of organic compounds of sulfur, oxygen and nitrogen, plus inorganic minerals such as sodium aluminum silicates or clays, quartz (SiO_2), and calcite ($CaCO_3$). Didn't know coal was so complicated, did you? One important and undesirable mineral found in coal is pyrite—iron sulfide, FeS_2. When coal burns (oxidizes), the pyrite is also oxidized:

$$4\,FeS_2 + 11\,O_2 \rightarrow 2\,Fe_2O_3 + 8\,SO_2$$

So burning coal produces sulfur dioxide and even some small amounts of sulfur trioxide, which, when dissolved in water, is sulfuric acid! Ferric oxide, Fe_2O_3, is only one of several constituents in the finely-divided fly ash, or particulate matter, or soot, formed when coal burns.

And then there's petroleum or crude oil, a liquid mineral found in the ground in many parts of the world. The word petroleum comes from the Latin, meaning rock oil. The United States, the Soviet Union and some Middle Eastern countries are the world's leading producers. Petroleum is predominantly a mixture of hydrocarbons—compounds of carbon and hydrogen—but like coal it also contains some organic compounds of sulfur, oxygen and nitrogen. A typical one of these compounds has the formula $C_{13}H_{10}S$; chemists call it methyldibenzothiophene for short. *If I write this on the board they'll groan and think they have to memorize it for an exam. If I don't, they'll get all shook up. You can't win.*

When this compound burns,

$$2\,C_{13}H_{10}S + 33\,O_2 \rightarrow 26\,CO_2 + 10\,H_2O + 2\,SO_2,$$

carbon dioxide, water and sulfur dioxide go off into the air. While coal and petroleum don't contain much of such compounds, amounting on the average only to a few percent of sulfur by weight, the several hundred million tons of coal and oil that we burn each year put more than 25 million tons of SO_2 into the air, along with millions of tons of particulate matter.

Gasoline is a product made from petroleum which we also burn a lot of. When gasoline is burned in the internal combustion engine of an automobile, we get another pollutant: carbon monoxide, CO. Carbon monoxide is produced whenever we have incomplete combustion. (Another source of carbon monoxide for many people, incidentally, is tobacco smoke. Smoking is banned in many public places nowadays, partly because it isn't healthy to have a high concentration of carbon monoxide in closed spaces. The reason for this is that carbon monoxide combines with the hemoglobin in the blood, decreasing the capacity of the blood's red cells to transport oxygen.)

Let's take $C_{10}H_{22}$ as a typical hydrocarbon compound in gasoline. When that burns in an engine with no pollution control devices attached, about 10 percent of the carbon ends up as carbon monoxide:

$$C_{10}H_{22} + 15\,O_2 \rightarrow 9\,CO_2 + CO + 11\,H_2O.$$

The Environmental Protection Agency (EPA) estimates that in 1971, just from our hundred million cars, trucks and airplanes, we put about 78 million tons of carbon monoxide into the air in this way!

A fourth air pollutant which is formed when any fuel is burned is due not to the fuel itself but to any high-temperature flame burning in the air. At high temperatures, there can be a reaction between the nitrogen and oxygen in the air which produces nitric oxide:

$$N_2 + O_2 \rightarrow 2\,NO,$$

and at lower temperatures, the NO can react with more oxygen to form nitrogen dioxide, NO_2:

$$2\,NO + O_2 \rightarrow 2\,NO_2$$

Thus, burning any fuel in air can produce NO_x. EPA estimates that we put about 22 millions of tons of NO_x into our air in this way in 1971. *Wow! I've put six equations on the board already. This is getting a little too heavy; they're starting to look at the clock. I'd better shift gears and start tying it all down for them.*

Well, in a very small nutshell that might begin to give you an idea of where our energy-mineral-pollution problems come from. Experts familiar with the energy situation predict that for at least the next 50 years we'll still be burning huge quantities of fossil fuels to get the energy we need to keep our industrialized and urbanized society going. By the year 2000, we expect to get significant amounts of energy from nuclear power, but that has its own set of problems. In particular, the safe storage of radioactive waste materials will be crucial to any widespread use of nuclear power plants.

In order to give you a better feeling for the facts and problems associated with the use of energy and other resources, I've selected an assortment of articles for you to read at home. You can pick up a package of these articles on your way out at the end of the lecture. First there's a very interesting historial review of how new energy supplies have influenced the way in which our society has developed since man first discovered fire—the first energy source he could manipulate to his own benefit. The author, Dr. Melvin Kranzberg, is a historian: Callaway Professor of the History of Technology at the Georgia Institute of Technology. He's also editor-in-chief of *Technology and Culture,* a journal that prints articles on the history of technology and its relation to society. (Yes, there are such magazines.) And as if that weren't enough, Dr. Kranzberg is chairman of the Historical Advisory Committee of NASA. (The purpose of this Committee is to see that the history of America's efforts and accomplishments in space are accurately recorded while they are still fresh in people's minds.) So he is certainly qualified to tell us how we got into this energy and mineral mess in the first place.

The next three articles in the package discuss some of the possible solutions to our energy problems: alternatives to the burning of fossil fuels. Of course, there are dozens of possibilities. People are talking about windmills, harnessing the ocean's tides and waves, solar power, tapping the heat of the earth (geothermal power), and designing automobiles powered by everything but pigeons. But I've just picked out three interesting possibilities for you to read about.

First is an article on nuclear power. Nuclear power has been a hoped-for solution to man's energy problem for more than thirty years, but it is only now finally beginning to come of age as a source of significant amounts of energy. A brief review of the status of nuclear power plants is given in the article, reprinted from *Newsweek* magazine.

Besides looking for new ways of producing energy, we could try to find more efficient ways of using the methods we already have. Magnetohy-

drodynamics, or MHD, while still in an experimental stage, has the potential of being a more efficient way to generate electricity than either conventional or nuclear power plants, because of the fact that MHD generators operate at much higher temperatures. The article by Edward Edelson from *Popular Science* magazine explains how MHD power generation works. Edelson also discusses the possibility that MHD may prove to be a less polluting source of electric power than people previously expected. MHD research and development is one of the cooperative programs being worked on jointly by the Soviet Union and the U.S. as part of the general *detente* of the past few years.

A very intriguing way to increase our energy supply while solving another problem at the same time is to use waste materials, or trash, as a fuel. Sizable amounts of iron and other metals, plastics and glass can be recovered from this process, thereby easing our mineral problem while generating "Trash Power," as discussed in the article of that title by Marjorie Mandel.

Next, the articles in your package turn to the problem of air pollution. To orient you, I've included first a couple of diagrams which show what air pollution consists of and where it comes from. I've also thrown in a diagram which summarizes the whole U. S. energy balance picture: how all our energy is produced and what it's used for.

Following these diagrams is an article on "Air Pollution and Human Health," taken from the annual report of an organization called Resources for the Future, Inc. It is based on a report written by Drs. Lester B. Lave and Eugene P. Seskin, two economists—not scientists—in the Graduate School of Industrial Administration at Carnegie-Mellon University. The deep involvement of economists shows you that energy and pollution are far from being only technological problems.

Two more articles complete the package: one on mineral resources and one on that most important resource of all, food. The mineral article is a detailed review of the mineral supply situation in the United States. More attention to this subject is certainly needed, and the author, Dr. Morgan, who is Assistant Director of Mineral Position Analysis at the Bureau of Mines of the U. S. Department of the Interior, is well qualified to provide the data we're going to need as a basis for future decisions.

And finally the last article, reprinted from *Newsweek,* calls attention to the critically important food supply situation in both this country and abroad. It speaks for itself.

After you've read these articles, we'll see how scientists, engineers, economists, industrialists, politicans and an informed public working together can help us to understand—and perhaps solve—these serious problems.

Class Dismissed.

Irving Wender
Bernard D. Blaustein

1
ENERGY, TECHNOLOGY AND THE STORY OF MAN

MELVIN KRANZBERG

The physicists have defined energy as the ability to do work, and they have postulated thermodynamic laws relating different forms of energy to one another and to work performed. But energy is more than a physical phenomenon: It is a social phenomenon. The way in which energy is produced, controlled, and applied—used and misused—helps determine the nature of society. Or, as Fred Cottrell, the sociologist, stated, "The energy available to man limits what he *can* do and influences what he *will* do."* Certainly man's material civilization—and much of his cultural life—is dependent upon a technological base, which in turn rests upon man's use of energy. As man has learned to control and apply energy in different ways, society has undergone concomitant changes.

Our prehuman forebears might never have evolved into our present species had it not been for their ability to control energy in the form of fire. During the many climatic changes which occurred during the eons of geologic time, prehistoric men might not have survived without fire to keep warm. It is not surprising that fire was regarded as sacred—a gift from the gods, or, as Greek legend had it, a theft from the gods by Prometheus.

Fire also improved man's chances for survival by increasing his food supply. By enabling him to live in colder areas, fire enlarged the territorial range of man's food-gathering activities. It also made possible cooking, allowing unpalatable or indigestible foodstuffs to be converted into assimilable human energy.

Fire extended man's use of materials. Early man already used stone implements and tools, but the development of pyrometallurgy, enabling him to use the store of metals locked in the earth, greatly extended his power and skills. Copper, bronze, and, later, iron provided man with materials which he could utilize to control and subdue nature, as well as his neighbor.

*Cottrell, W. F: *Energy and Society.* Greenwood Press, Inc., Westport, Conn., 1955. p. 2.

From *The Science Teacher,* March 1972. Reprinted with permission.

Yet even with the application of fire, the greater part of energy available to man for performing work came mainly from his own strength. There were two ways by which man could augment his muscle power. One was by domesticating animals, so that they would perform work for him; the other was by devising tools and implements which would amplify and extend the power of his muscles. An increase in energy through the domestication of animals, about the seventh millennium BC, could not make itself felt until more efficient use could be made of animal muscle power by the invention of the wheel (about 2500 BC) and, much later, of a more efficient harness for horses. The increase in power through tools had come much earlier.

One of the first "machines" was the bow-and-arrow, which might be viewed as a machine for storing energy and releasing it in directed fashion. When the bowstring is slowly drawn back, human muscle power transmits energy to it; this energy is released suddenly when the archer shoots. The bow-and-arrow and other weapons expanded food supply by enabling man to kill small game at a distance. Other early tools and devices also multiplied human muscle power; this was especially true of the wheel, which made it easier to transport heavy loads over long distances. By classical antiquity, Archimedes could classify the "five simple machines" and analyze the mechanical advantage which they gave man in manipulating things.

Even with only rudimentary tools, human muscle power can perform feats of great magnitude if organized effectively. The chief example is the pyramids. Although built before the Egyptians had the wheel, the pyramids demonstrate the prodigious accomplishments of human muscle power when effectively marshalled in the performance of a collective task.

It has been claimed that slavery militated against more effective use of human energy. When faced with a problem requiring the exertion of more force, the ancient engineer simply added more slaves to the work gang instead of devising some ingenious mechanical solution which would lessen the strain on human muscles. Proof of this is adduced from the fact that the water wheel was known in Roman times, but Romans scarcely made use of it. Instead, they continued to rely on human and animal muscle power.

The Middle Ages witnessed a veritable "power revolution." Rome's decline coincided with the decline of slavery and stimulated the application of new power sources. Water wheels came into widespread use and were improved. The vertical water wheel, introduced in the fourth century after Christ, for instance, had a power capacity of 2 kilowatts, compared with only three-tenths of a kilowatt for the earlier horizontal type.* By the time of the Domesday Book (1086), there were some 5,000 mills in England, amounting to one mill to every 400 of the population. First used for grinding grain, water power was later applied to a great diversity of industrial uses, most importantly to drive the bellows of blast furnaces so that the economical process of casting metals could come into widespread use.

Another source of inorganic power contributed to the medieval power revolution: wind. Although wind had been used to drive sailing vessels since antiquity, its use as a power source in the West dates from about the twelfth century. Windmills provided power in flatlands where the fall of streams was too slight for a water mill and where a mill dam would flood too much land.†

The medieval power revolution also enlarged animal power by improvement of the harness, which had remained unchanged since about 3000 B.C. The old harness had been held in place by a strap around the neck; as soon as the horse exerted a heavy pull, it choked itself. Furthermore, the ancients did not know how to harness horses in file in order to multiply their tractive power; nor did they use horseshoes, and their horses often suffered foot injuries on rough terrain. The rigid horsecollar, probably introduced from Asia during the early Middle Ages, together with horseshoes (ca. the tenth century) and the tandem harness, multiplied the effective pulling power of horses by some three to four times over that in antiquity.‡ A horse driving a machine with the new and more efficient harness was the equivalent of 10 slaves, while a good water wheel

---

*Starr, Chauncey: Energy and Power. *Scientific American 225*:37–49, 1971.

†White L., Jr. "Medieval Roots of Modern Technology." In *Perspectives in Medieval History*. K. F. Drew and F. S. Lear, Editors. University of Chicago Press, Chicago, Illinois. 1963.

‡White, L., Jr.: *Medieval Technology and Social Change*, Oxford University Press, New York. 1962, pp. 57–69.

Belmont Nail Works, Wheeling, W. Virginia, with Ohio River in foreground.
(From Illustrated Atlas of the Upper Ohio River, 1877).

or windmill provided the work of up to 100 slaves.

The medieval power revolution, coupled with technological innovations in agriculture and machines, laid foundations for the renewal of town life and the beginnings of industrial technology. The Renaissance saw continued growth in the exploitation of water and wind power as power devices increased in size. By the seventeenth century, the Marly works which pumped water for Versailles had a power output of 56 kilowatts; similarly, the capacity of windmills grew from several kilowatts to as much as 12 kilowatts. In addition, the Renaissance developed complex gearing arrangements for the transmission of power from water wheels and windmills so that more efficient use could be made of the energy input.*

By the eighteenth century Europe was in desperate need of new power sources. Windmills were effective only in flatlands, such as the Netherlands, where the terrain did not interfere with a steady wind. Water power was intermittent in operation; the flow of water would decline during dry seasons or freeze in cold weather. In England, all the good industrial sites—that is, where there was a sufficient flow and fall of water—were already taken up by factories crowded close together. The need for a new source of power was especially prevalent in the mining areas, to pump water from mines. The steam engine answered that need and became the characteristic power source of the Industrial Revolution.

The steam engine, and the Industrial Revolution of which it was a part, completely transformed the economic, social, and cultural life of Western Civilization—and ultimately of the entire world. For almost all of human history, the hearth and home had been the center of production, and agricul-

*Keller, A. G., *A Theatre of Machines*. The Macmillan Company, New York, 1964.

Drawing of the first American railway train as it appeared in July 1832 on the Mohawk and Hudson Railway.

ture had been the chief occupation of the vast bulk of mankind. Men lived in rural areas, their horizons limited, and with a standard of living scarcely above subsistence. Industrialization changed all that. The factory became the center of production; the city became the center of human life and production; family relationships changed, while traditional institutions, such as religion, lost their hold upon men's minds in the new urban surroundings.

A transformation also occurred in the centers of political power. The steam engine gave the industrial advantage to countries with abundant supplies of coal. This advantage was reinforced when steel became the basic industrial material. Because the making of steel required more coal than iron, the location of the energy resource determined the center of steel production. It is not surprising that Britain, with its iron and coal deposits, became the world's industrial leader and the dominant political power during the nineteenth century.

The steam engine was the first mechanical prime mover to provide mobility. Through the steamboat and locomotive, energy from steam revolutionized transportation and brought the world closer together.

Steam both epitomized and embodied the fundamental technological change of the modern era—an incredible jump in available energy. Before the age of steam, the sum total of energy which man could effectively convert to his purposes through wind and water, through animal and human power, was quite limited. Then, in the nineteenth century, man liberated the power of fossil coal through the steam engine on a scale never before possible. The new power, multiplied or divided almost at will, was applied to uncounted tasks.

Yet the steam engine had some disadvantages. It was heavy and cumbersome; almost a century and a half after its development, a reciprocating steam engine could operate at only 23 percent of thermal efficiency. Nevertheless, the power revolution ushered in by the steam engine expanded under the impetus of scientific discovery and technological innovation which created a new prime mover, the internal-combustion engine, and a new form of energy, electricity.

The internal-combustion engine enlarged, accelerated, and altered the social changes already occurring as a result of steam power. Small, light, and powerful, the internal-combustion engine personalized the amount of power available for each individual, provided transportation for everyone, and made readily available a source of power with which to do a number of tasks, including fulfillment of man's ancient dream of flying. The automobile's effect on American society is so marked as to require no recounting here. The mobility of American society is evident in every facet of our daily life—and this is the result of energy for transportation made available to everyone through the automobile. Furthermore, the automobile represents a major factor, perhaps *the* major factor, in the American economy: One out of every eight workingmen in the United States is employed in a task directly connected with the automotive industry.

Equally portentous was the discovery, development, and utilization of electricity. A single comparison with the past illustrates one dramatic change wrought by electricity. One of the romantic episodes of the American West was the Pony Express, begun in 1860, but it was only the ultimate exploitation of a form of communication centuries old—a message carried by a man on a horse. The trip from St. Joseph, Missouri, to San Francisco took ten and a half days—ten fewer than the best stagecoach time. A year later, however, the transcontinental telegraph was completed, and the time for a message to reach the Pacific was again cut, this time to a fraction of a second. Within another 15 years the telephone was a crude though working reality, and the Atlantic Cable shortly after mid-century brought messages across the ocean in an instant. Today, through telecommunications satellites, events occurring halfway across the world are flashed immediately onto television screens in our living rooms.

Electricity has become so much a part of our daily lives that we sometimes fail to recognize how unique and important it is—until a storm or malfunction cuts off our power. We fail also to realize how really new it is in terms of the long time span of human history. Yet it was less than a century ago (1882) that the first electric power station for private consumers, the Pearl Street Station, was established by the Edison Company in New York City. It served a modest load of approximately 1,400 lamps, each taking about 83 watts and constituting an electric demand of 33 kilowatts.

Electrical energy found still another use in the twentieth century. In the census of 1890, the United States Bureau of the Census introduced machines to sort and tabulate data on punched cards, providing the prototype of new means for storing and manipulating information. Previous applications of energy through mechanical devices had lifted the burdens off men's backs. Now the use of energy for information purposes began lifting the burdens off men's minds, freeing them from dull and repetitive tasks.

And, just as the development of earlier energy sources had allowed the exploitation of a wider field of materials, so did the new electronic devices allow for manipulation and exact control of the productive process. Automation reduced the need for human labor while increasing and standardizing output. For the first time in human history, a society of abundance and affluence was possible. The implications of this for work and leisure are still not completely understood.

Contemporaneous with the new applications of energy in the twentieth century—and, indeed, dependent upon them—was a revolution in agriculture. Farm mechanization demonstrates how the application of energy to food production has eased man's toil and, at the same time, increased productivity. As late as half a century ago, approximately one-quarter of the farm acreage in the United States grew crops for feeding the 25 million farm horses and mules; that acreage has been freed to provide food for human beings.

The growing utilization of energy on the American farm has meant that there is less need for human and animal muscle power. Rural workers have been displaced, for they are no longer needed to till the soil; the family farm is fast disappearing, and agro-business is taking its place. A vast migration from rural regions to cities has accompanied the growth of farm productivity, and this has given rise to urban and also racial problems as southern Blacks have migrated to northern industrial centers.

In brief, the manifold applications of energy in production, both industrial and agricultural, have given rise to a new type of society where production, which had occupied so much of man's time and energies over the course of the centuries, no longer presents a problem.* The census figures reveal the magnitude of the social changes. What was primarily still a rural and agrarian nation at the beginning of the twentieth century had by the middle of the century become predominantly urban. Late in the 1950s the number of people employed in the service sector of the economy surpassed those engaged in production of goods.

Although ushered in with an act of destruction which still echoes in the conscience of mankind, a new form of energy—nuclear power—marks one of the greatest triumphs of science and technology in our time. Radioactivity had been discovered near the close of the nineteenth century and had caused excitement in a small segment of the scientific community, but the idea of harnessing this energy for useful purposes was imagined only by science fiction writers. In *A World Set Free,* written in 1914, H. G. Wells predicted that nuclear energy would be invented about 1940 and, most presciently, that it would be used in war about 1950. Wells thought that about the middle of the 1950s, the world would come to its senses and realize the possibilities offered by nuclear energy in the form of unlimited, cheap, and ubiquitous energy which would, in a very real sense, set men free from the limitations of previous energy sources.†

But something happened on the way to Wells' Utopia. Neither man nor energy is yet free. What is worse, the country is confronted with an energy "crisis." On June 4, 1971, the President of the United States sent a message to Congress acknowledging that the country had entered a period of increasing demand for energy and of growing problems of energy supplies. Scientific and technological factors account partially for the energy crisis, but much of the difficulty has arisen from social factors, especially public hostility stemming from fears of thermal pollution

*W. W. Rostow calls this "the age of high mass-consumption." In *The Stages of Economic Growth.* Cambridge University Press, Cambridge,. England. 1960. Chapter 6.

†Similar possibilities are set forth in a more recent work by Glenn T. Seaborg and William R. Corliss: *Man and Atom: Shaping a New World Through Nuclear Technology.* E. P. Dutton & Co., Inc., New York, 1971.

and radiation hazards and alarm over danger to the environment from older forms of energy applications.

Americans are demanding a quality environment. They are appalled at ugly strip mines, oil slicks from tanker spills and leaky offshore wells, denuded corridors of land for energy transmission lines, sulfur oxides and fly ash from power plants, noxious emissions from automobile exhausts, and the real or imaginary specter of radioactive perils from nuclear centers. Scientists and technologists engaged in the production and application of energy are thus faced with a major problem: how to protect the environment and still provide for the "good life."

That question poses basic issues about values. The value scheme implicit in our past profligate use of energy was based upon a human history of scarcity—scarcity of material goods to satisfy basic needs of food, clothing, and shelter. Now that we have the potentiality for a sufficiency of material goods, the environmental issue has become of overwhelming importance. Heretofore, American society, with its orientation toward material development, has always considered economic growth as essential to the nation's well-being. Yet if proper cognizance is taken of environmental factors, we might have to slow the pace of economic growth or change its direction.

What message, then, does the story of the development of energy and its impact on man's history have for us? It tells us of the great opportunities offered by human imagination and ingenuity in converting energy to man's use and how such uses inevitably affect our social destiny. With this knowledge in mind, we can face with confidence the current issues involving energy production and application. We can examine our values, we can decide how we want to live, and we can have due regard for the other peoples of the world and for future generations within our own country. We can then arrive at a comprehensive national policy which will take our resources, our science and technology, and our aspirations into account. The present "energy crisis" does not bring man into confrontation with his science and technology; it forces man to confront himself.

2 THE NUCLEAR SPEED-UP

THE STAFF OF *NEWSWEEK* MAGAZINE

There seems no end to the variety of crash programs envisioned by energy experts to mitigate the energy crisis: solar power, wind power, even garbage power (by conversion of wastes into oil) are currently being explored. But the nation's $3 billion nuclear-energy program is almost certainly the most viable long-term answer to the problem—the key question is how fast it can be speeded up.

Already, the 38 nuclear reactors now operating in the U.S. are producing more than 5 per cent of the nation's electrical power—a fivefold increase over their output in 1970. According to estimates made by the nuclear industry before [the President's] call for an acceleration of nuclear development, some 140 reactors will be supplying more than 20 per cent of American electrical demand by 1980. After that, the proportion of nuclear-generated power should increase apace, to about 45 per cent of national electricity usage by 1990 and to 60 per cent by the end of the century.

In order to accelerate this timetable, scientists will have to work on at least three fronts at once because there are three basic types of nuclear reactors, each at a different stage of development and each posing its own problems:

LIGHT-WATER REACTORS

The light-water reactor, the type now in general operation, uses as its fuel the form of uranium known as uranium-235. Because this form makes up only 0.71 per cent of naturally occurring uranium, the fuel for light-water reactors must be prepared by an expensive process of enrichment, involving either large centrifuges or complex separation of gases.

The reactor works in this way: rods of enriched uranium are placed in the center of the reactor, known as the core, separated by rods of a material such as graphite, which is known as a moderator. In action, each atom of uranium-235 produces two or three neutrons when it splits up; these neutrons then

From *Newsweek,* 10 December 1973. Copyright Newsweek, Inc. 1973. Reprinted by permission.

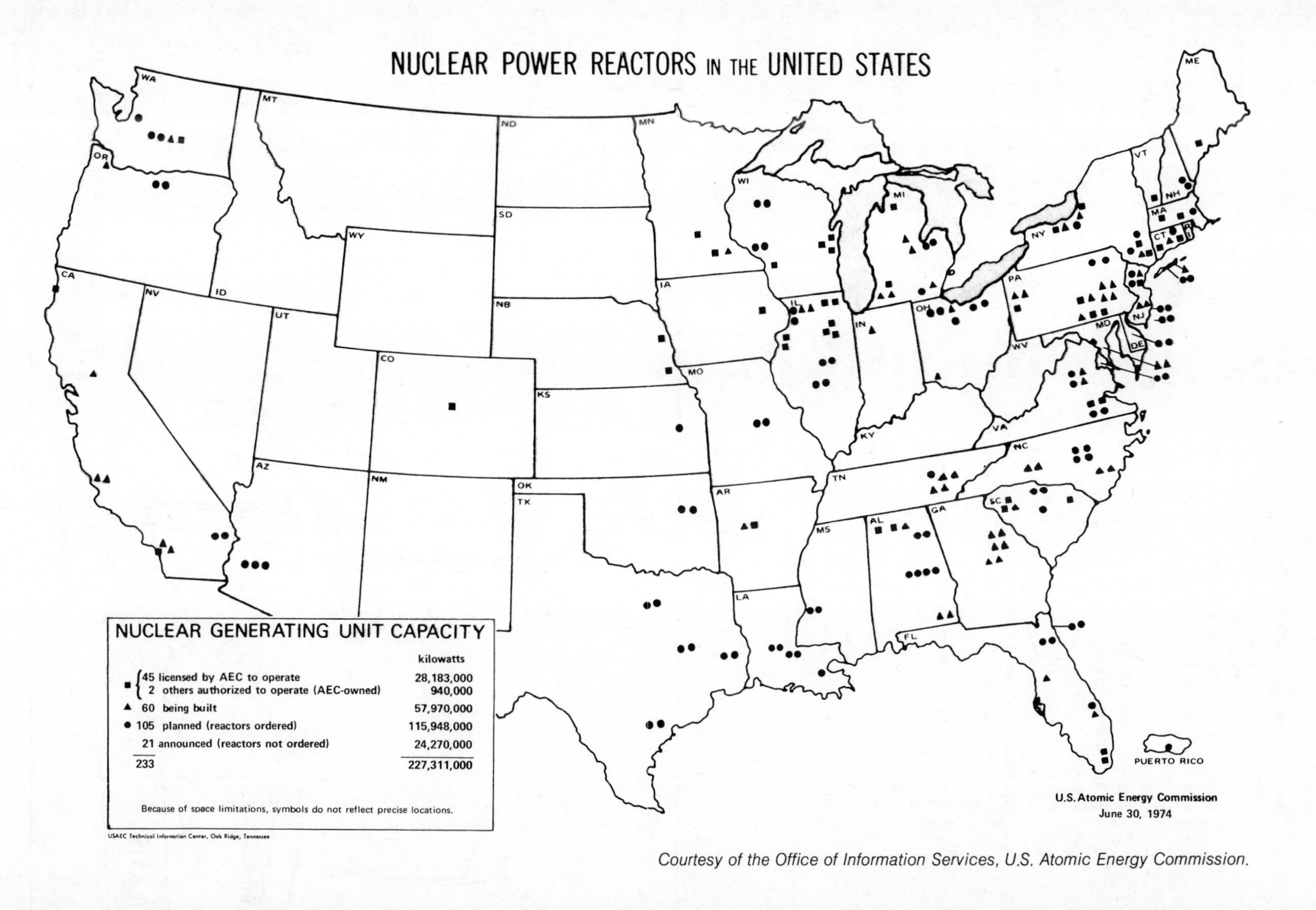

Courtesy of the Office of Information Services, U.S. Atomic Energy Commission.

strike more uranium-235 atoms, splitting them and in turn producing more neutrons. If allowed to continue in this way, the chain reaction would run out of control like an atom bomb. However, the moderators between the uranium fuel rods absorb some of the stray neutrons, thus keeping the reaction within bounds. And for fine tuning of the rate at which the reaction proceeds, the reactor is equipped with "control rods," generally made of some other material that absorbs neutrons, which can be lowered into the core or withdrawn from it as necessary.

When the reaction is proceeding at the correct rate, it yields a great deal of heat. This is removed from the core by a coolant—generally water under pressure—which heats up as it passes through the core. The water is then used outside the core to heat more water to produce steam, which in turn drives a turbine to generate electricity.

The nuclear industry regards these reactors as clean, safe and efficient. However, they have run into a number of problems during the start of operation, and nuclear experts concede that it takes about two years to overcome all the teething troubles of each new reactor. Environmental critics say the problems could portend major accidents. At present, none of the six reactors in New England is operating at full power. Three of them are restricted by a vibration problem. But the nuclear industry says that the trouble is now solvable.

For the moment, what the nuclear industry itself sees as the major problem with the light-water reactors is how to reduce the present ten-year time span that comes between an order for a new plant and the point at which it actually begins to produce electrical power. If this lead time could be shortened to five or six years, as [the President] suggested, then nuclear reactors could be built just as quickly as coal-fired power plants.

At present, say both the manufacturers and AEC officials, the U.S. is producing what is in effect a Rolls-Royce line of nuclear-power plants. Each one is custom-made, custom-approved and custom-regulated. What the energy crisis requires, these experts insist, is a complete retooling of the

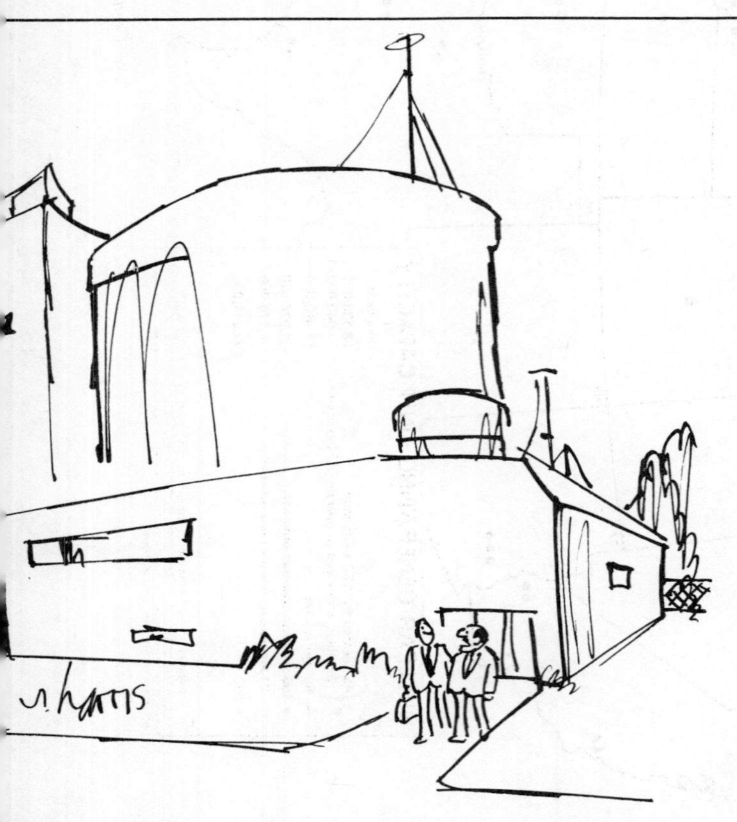

"Remember—it's better to light just one little nuclear power station than to curse the darkness."

From Sidney Harris, American Scientist.

entire system, a program that, in effect, would permit the construction of nuclear-power plants on an assembly-line basis.

For its part, the AEC is now actively encouraging mass production and mass approval of plants. According to one plan now under consideration in Washington, major manufacturers of reactors would submit to the agency final plant designs that, once approved, could be immediately ordered by utilities anywhere in the country. Another speed-up scheme envisions blanket approval of "designated sites" for future construction of unspecified reactors.

"If we can get designated siting, we can save a year and a half for each reactor," explains AEC commissioner William Doub. He adds that standardized design of reactors could save another eighteen months to two years. Over-all, he says, the two approaches could hasten increased nuclear-power production by three years or more—"pretty close to the President's objective."

BREEDER REACTORS

By contrast with the light-water reactor, the breeder reactor uses the most common form of uranium, known as uranium-238, in addition to uranium-235. This is possible because the reactor operates at much higher temperatures. At such temperatures, not only does the fission reaction involving uranium-235 take place, but uranium-238 atoms also react with neutrons to produce atoms of yet another fissionable material, plutonium-239. The over-all reaction produces more fissionable material than it starts with—hence the name breeder reactor.

At present, the science of the breeder reaction is well understood, but its translation into commercial power-producing equipment is still in the future. The major problem with breeders involves removing the heat from their cores. Because of the high temperatures, liquid metals such as sodium are preferred to water for this task. However, producing leakproof plumbing for this liquid metal is an awesome task, and many critics of the breeder doubt that the nuclear industry will be able to provide a safe means of tapping its huge power. Environmentalists are also concerned at the prospect of transporting and storing the large amounts of fissionable plutonium-239 that will accumulate in the reactors. The AEC, however, remains confident that it can overcome the problems and thus stretch the nation's supply of uranium much farther than is possible with light-water reactors. If no major technical snags arise, breeder reactors could be generating large amounts of electrical power by the 1990's.

FUSION REACTORS

In the same way that light-water and breeder reactors represent controlled versions of the atom bomb, the fusion device is designed to mimic the hydrogen bomb in slow motion. The major fuels for fusion are deuterium and tritium—two heavy forms of hydrogen. Deuterium is abundant in sea water, and tritium is a man-made isotope. At temperatures of millions of degrees, these elements fuse together to produce helium atoms, releasing vast amounts of energy.

From a scientific point of view, the problem is to keep the reaction going long enough to produce a net gain of energy, for simply starting the process requires huge amounts of power. The key to achieving this is to contain a very dense mixture of the elements within a very small space while it is heated up to the appropriate temperature. The major approaches used to achieve this goal have been containment of the mixture inside strong magnetic fields and the instant induction of very high temperatures using unusually powerful lasers. To date, neither technique has proved successful, and even the most optimistic estimates put commercial use of fusion no closer than the year 2020.

For the immediate future, then, it is clear that the highest priority probably will go to getting standardized light-water reactors on line as quickly as possible, while research on the breeder and fusion reactors proceeds apace. The environmentalists will continue to oppose this, but [in December of 1973] AEC chairman Dixie Lee Ray herself took out after the environmentalist critics in an unusually sharp attack. "They have used innuendo and inaccuracies to build a case against nuclear power largely on emotional grounds," she said. "We do not believe that the people will be fooled." . . .

3

MHD GENERATORS:

Super Blowtorches Deliver More Power With Less Fuel

EDWARD EDELSON

Standing inconspicuously in the parking lot behind the Avco Everett Research Laboratory near Boston is a tarp-covered square tube, about as long as a family car and as wide as a desk, that's worth a second look. It's the heart of the Mark V, a magnetohydrodynamic generator whose 32-megawatt output still holds the world's record. The Mark V was dismantled a few years ago, but, says Stanley Perry, an Avco engineer who spent a lot of time with it, "I just didn't have the heart to get rid of it."

The Mark V is all too symbolic of the MHD story to date: It's been around for a while, it's achieved tantalizingly promising results—and no one knows what to do with it.

For years, a handful of MHD proponents—most notably Arthur Kantrowitz of Avco—have been saying that their system could increase the efficiency of electric generating plants by 50 percent or more. And for years, MHD research has been limping along on a minimal budget: a few dollars from a corporation here, a small grant from a utility there, a minor effort on the part of the federal government.

There have been some encouraging signs for MHD lately, such as a U.S.-Soviet agreement to cooperate in research, and a slight increase in federal appropriations, but the energy crisis has created a moment of truth for MHD. As far as people in the field can tell, it's now or never. Either the government comes through with a quantum-jump increase over [1973's] $5.75 million, or the immediate future for MHD development looks dim indeed.

"If the energy crisis can't put MHD over," said one expert I talked to, "then nothing can."

MHD engineers have an offer they think you can't refuse: Give them the money and the time—about a half-billion dollars and a decade—and they'll give you a generating station that has no moving parts, is almost pollution-free, and produces perhaps 4500 kilowatt-hours of electricity for every ton of coal, compared to just 3000 kilowatt-hours for todays' best generating plants.

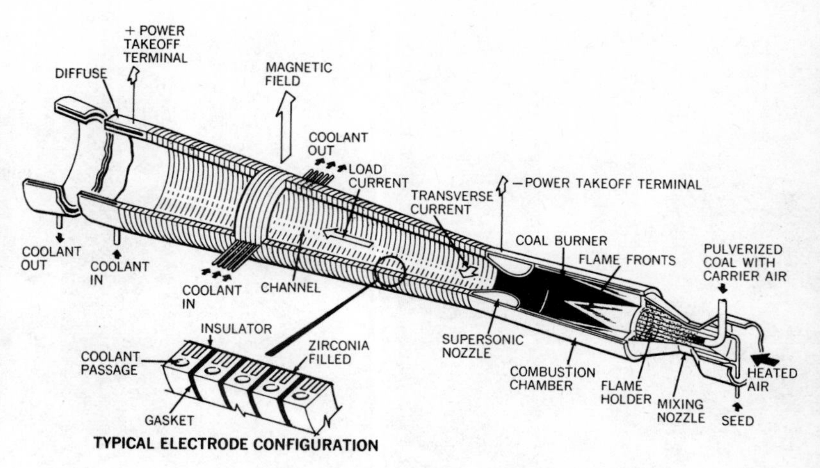

How MHD generator produces electricity: Pulverized coal is burned with preheated air to produce 5000-deg. gas. Metal "seed" particles added to the gas increase its conductivity. Gas shoots through rocket-like nozzle into generating channel at supersonic speed. Electrodes in channel are water-cooled and ceramic-coated to withstand high temperature and corrosion. In working MHD plant, generating channel would be within field of giant superconducting magnet.

For a nation that's suddenly fuel-hungry, that offer should be irresistible. But the small band of MHD enthusiasts have said the same thing for a long time without getting anywhere—and you can't blame the jaw-breaking name they're stuck with. If you look at the sort of problems MHD has to overcome, you'll understand why they encounter skepticism.

HOW MHD DOES IT

Magnetohydrodynamics is an energy conversion system—a way of turning heat into electricity by using a hot, conducting gas.

To generate electricity the MHD way, you heat a gas to about 5000 degrees, some five times hotter than in a conventional steam turbine. This turns it into what physicists call a plasma. That means that the electrons have been stripped away from many atoms so that the gas becomes a good conductor of electricity. Then you expand the plasma through a rocket-like nozzle, add a metal "seed" like potassium or cesium to increase the gas's conductivity still further, and send it roaring down an electrode-lined channel at about the speed of sound.

You have to protect the walls of that channel, and the electrodes through which electricity is drawn off, from all sorts of damage—melting due to heat, erosion caused by the sheer speed of the gas, corrosion from the seed metal atoms and, if you're burning coal, the impact of ash particles. What's more, you've got to put this channel inside the poles of a superconducting magnet whose field is 50,000 gauss or more and whose size approaches that of the family garage.

Downstream, you have to deal with the extra-high levels of eye-stinging nitrogen oxides produced by the high temperatures of MHD, as well as the sulfur oxides and ash that come out of any plant that burns coal or oil.

And you lose the extra efficiency promised by MHD if you cut corners anywhere. Lower gas temperature means less conductivity; a reduced electromagnetic field means less generating capacity; too much pollution means a visit from an inspector. So meeting the standards for MHD success seems like a tall order—and it seems even taller when you learn how limited MHD experience has been.

MHD theories have been around since Michael Faraday discovered the principle of the generator. But no one actually made an MHD generator go until after World War II. Even today, operating experience of these generators is not very impressive.

Engineers have run 10-kilowatt MHD generators for a few-hundred hours at a time. But operating time of megawatt-level MHD generators is measured in minutes. Mark V never even ran for an hour continuously.

THE SOVIETS' GENERATOR

The world's current largest plant is the Soviet U-25, designed for an ultimate capacity of 25 megawatts. Built in a northern suburb of Moscow, it has only a few hours' total experience at four megawatts or less. It does, however, have the distinction of being the world's only MHD generator that is tied into a working power grid. U.S. experts say the Soviets have made a major achievement

Soviet U-25 MHD generator, with potential capacity of 25 megawatts, is the world's largest, and is the only one tied to a working power grid.

in overcoming the engineering problems that accompany putting a promising concept into actual operation. Thus these experts are impressed with the U-25's interconnecting equipment, but not with the MHD core. For, one of them said, the Soviets went ahead and built the kind of MHD plant we could have had if research had ended in the mid-1960's.

Still, MHD people say they can talk realistically about a plant that would operate at 1000 megawatts for hundreds or thousands of hours. Their reasoning: Research has already overcome the critical problems, and the rest is just a matter of tinkering with known principles.

"The problems we face aren't of a basic physics nature," says Raymond Janney of Avco. "They're purely of an engineering nature."

The best reason for listening to statements like this is the hope that MHD can stretch our precious supplies of fossil fuels.

Today's best coal- or oil-burning plants, operating at 1000 degrees, turn about 40 percent of the fuel's energy into electricity. The rest is waste heat. Nuclear plants, running a few-hundred degrees cooler, are only about 33-percent efficient. MHD people talk about total efficiency in the area of 60 percent, meaning they'll get half again as much usable energy out of coal or oil as we do right now.

MHD would get the extra efficiency in two ways—by using its combustion gas directly, eliminating the waste of making steam to turn a generator; and by using a hotter gas—the basic rule of a heat engine says that the hotter you can run it, the more efficient it will be.

THREE MHD CONCEPTS

There are three MHD concepts aimed at reaching this promised land of high thermal efficiency. Far in the lead is open-cycle MHD, in which gases heated in a combustion chamber are sent directly through the generating channel. Much more work has been done on the open-cycle concept than on the competing closed-cycle proposals, in which heat is transferred either to a noble gas or to a liquid metal, which then runs through the channel. It's the open-cycle people, led primarily by Avco engineers, who are talking about building a pilot plant now.

But to believe that open-cycle MHD can make it, you have to look at a lot of small bits of evidence from several laboratories.

Take what looks like the key challenge in MHD—preserving the generating channel from the assault of that powerful blast of hot gas. Even today's best refractory ceramics, such as zirconia or magnesia, get ripped away too soon.

Solution: Feed in a refractory powder with the fuel so that there is continuous replacement of the coating that's scoured off. If the feed is just right, the channel gets the protection it needs.

Here, the MHD people think they've hit it lucky. Work at Avco and the University of Tennessee Space Institute indicates that ordinary coal slag does just as good a job of channel coating as the most exotic refractory material.

"If we had set out to create an ideal material for wall coating, we would have come up with something just like coal slag," exults Avco's Richard J. Rosa.

But that rejoicing comes after short-run tests; now the job is to show that coal slag will work in the long run. Avco's 500-kilowatt Mark VI is setting up for a 100-hour run, which could come this year. J. B. Dicks and his co-workers at the University of Tennessee are also aiming to run their MHD generator for 100 hours, feeding in pulverized coal ash all the time. The idea has looked good in brief runs, but as Dicks points out, "If we run it for a few-hundred hours, problems that we can't see now could become appreciable."

WHAT'S MISSING

As for controlling the amount of ash that goes in, the Bureau of Mines research center in Pittsburgh is working on a novel three-stage combustion chamber that would do just that. But it isn't ready yet.

Neither is the superconducting magnet needed for a large-scale generating plant. MHD researchers at Stanford University are about to get an 80,000-gauss superconducting magnet, but that will be used for basic scientific studies.

Without a superconducting magnet, the whole MHD generator idea would collapse, because an inordinate amount of energy would be needed for a conventional electromagnet of the necessary size. Yet, as Dicks points out, "No one has ever run an MHD generator with any superconducting magnet, aside from some very small experiments in Europe."

You can go through the proposed MHD plant component by component and hear the same thing. Consider the plan for controlling pollution. Nitrogen oxides, a special problem for MHD because they increase directly with combustion temperature, would be handled by a two-stage burning process, using an afterburner to eliminate them. Sulfur oxides and ash would be collected with the seed metal; the seed would be separated out chemically and recycled, while pollutants would be dumped harmlessly.

POLLUTION CONTROL IS INTEGRAL

The MHD advantage is that pollution control is not a tack-on as with today's systems. But, of course, none of the MHD pollution controls has been proven in prolonged plant testing.

And if open-cycle MHD has some open questions, the closed-cycle concepts are more like a gleam in someone's eye.

General Electric's Space Sciences Laboratory in Valley Forge, Pa., has something like a monopoly on closed-cycle noble gas MHD. Bert Zauderer, who heads the program, is enthusiastic about the scientific elegance of the idea: Instead of heating all the gas to 5000 degrees, the idea is to selectively excite only the electrons in the noble gas (in much the same way that electrons in a fluorescent bulb are excited), thus getting the same gas conductivity at a much lower overall gas teamperature.

As Zauderer pictures it, a closed-cycle, noble-gas MHD generator would have several columns of ceramic pebbles that would be heated by combustion gases and would transfer that heat to argon or another noble gas, which would then flow through the generating channel.

Zauderer admits that his scheme is five or six years behind the open-cycle concept, that it would need excruciatingly clean fuel to avoid fouling the ceramic pebbles, and that many scientists doubt that selective heating of electrons can be done. "We're the only ones who can get it to work," he said of his concept.

But even though Zauderer measures operating times in milliseconds—all his experiments have been done in shock tubes thus far—he maintains that given $25 million and eight or nine years, his closed-cycle scheme will reach demonstration-plant status. (GE is spending about $500,000 on MHD research this year.)

Still further behind is a closed-cycle MHD concept that would use liquid metal as the generating fluid. Michael Petrick of Argonne National Laboratory, where the closed-cycle liquid-metal work is being done, likes the idea because of its low operating temperatures—in the neighborhood of 1500 degrees—and because the system could be used with the liquid-metal fast-breeder reactor that's now under development and—perhaps—with fusion plants.

The problem with liquid metal—sodium is the leading choice—is that it doesn't flow fast enough, and engineers haven't been able until recently to manage to pump in gas to speed the flow. But just now, Petrick is talking about "something of a breakthrough," a method of injecting gas that has doubled efficiency. "It's beginning to look really attractive," he says.

MHD TIMETABLES

Petrick has a timetable: Run a 100-kilowatt generator this year, move to a one-megawatt generator in three or four years, have a 20-megawatt pilot plant in operation by 1982. That would cost about $45 million, and would have liquid-metal MHD ready when the first fast breeder is scheduled to go on line.

The open-cycle people have their own timetable, one that was presented to a congressional committee last year.

First, we keep on supporting basic MHD research at a level of about $8 million a year. This basic work would include such efforts as the conversion of a wind tunnel at the Air Force's Arnold Engineering Development Center in Tennessee into a 58-megawatt MHD generator, bigger than any to date, that would operate in 10-second bursts to test new ideas.

It might also include construction of a 20-megawatt channel by Avco for the Soviet U-25. The Soviets say that if we build the channel, they'll test it under operating conditions.

Second, we would start planning a 50-megawatt pilot plant right now, and would have it in operation by 1976. One year later, using data from the pilot plant, engineers would start planning a 1000-megawatt MHD plant, which would go on line by 1982.

As they envision it, the plant would consist of an MHD "topping section" that would use the open-cycle method to extract perhaps 20 percent of the fuel's energy. Spent gas from the MHD section would go to a conventional steam generator that would get another 40 percent of the energy out. Total operating efficiency: about 55 to 60 percent, subject to improvement with later models.

Total cost of the entire program is $560 million—which, MHD experts quickly point out, is about the price of one large conventional generating plant. If MHD works, the savings in fuel could run to billions. Just how many billions, no one knows, because the price of all fuels seems to be going up indefinitely.

Now there are only two questions to be answered: Will MHD work? And are we willing to put up the money to get the answer?

4

TRASH POWER—ELECTRICITY FROM RUBBISH

MARJORIE MANDEL

Trash is piling up all over the United States, millions and millions of tons of it a year. We are running out of places to dump it. Even the ocean, long used as a dump by coastal communities, no longer seems to be a bottomless sink.

And dumping seems more and more an uneconomic way of disposing of trash. Trash contains substances that can be burnt to provide energy and substances that can be recycled as raw material for manufacturing. A society faced with various shortages is beginning to look into its garbage cans for help. A large project of this sort is under way in St. Louis.

By mid-1977, the St. Louis metropolitan area plans to become the first region in the United States to use essentially all of its solid waste to generate electricity. Thus it hopes to solve its solid waste disposal problem, eliminate the need for incinerators and landfills and save millions of dollars in fuel costs.

St. Louis' Union Electric Company announced plans [in early 1974] for a system that will generate approximately six percent of the company's power from solid waste. The company anticipates that the system will be able to handle the 2.5 to 3 million tons of solid waste produced annually in the city of St. Louis and six adjoining Missouri and Illinois counties.

"The project is the real first of its kind," according to Union Electric President Charles Dougherty. It will be built without governmental subsidy, and the $70 million of private capital for the system and the $11 million annual operating costs will come from the heating value of the solid waste, sale of recyclable materials sorted from the waste before it is burned, and dumping fees. About two and a half pounds of solid waste are required to substitute for one pound of coal in the electrical generation process.

Union Electric's decision to develop a regional Solid Waste Utilization System follows its evaluation of an experimental prototype system, the Energy Recovery Project, which has used about 200 tons of city gar-

From *Science News, 105:* 212–213, 30 March 1974. Reprinted with permission.

Courtesy of Union Electric Company.

bage a day mixed with coal to generate electricity since mid-1972. The prototype was a joint project of Union Electric, the city of St. Louis and the Environmental Protection Agency (EPA). The $3.3 million prototype was funded by a $2.2 million EPA grant and matching local funds.

Under the plan, Union Electric will establish about half a dozen collection-transfer centers in the metropolitan area which will receive solid waste from private and public haulers and transfer it to closed containers for rail shipment to processing facilities.

From that point, the proposed system will operate much like a large-scale expansion of the prototype. At present the Energy Recovery Project processes about 200 tons of residential trash daily—about 30 per cent of the city's total—through a large hammermill. The new system will have a daily handling capacity of 7,000 to 8,000 tons and will most likely not be limited to residential trash. "There may conceivably be some industrial wastes we may want to take a second look at," Dougherty said.

After shredding in the hammermill, the shredded material is dropped down a large vertical wind tunnel, or air gravity separator. The heavier materials drop through an air jet onto a conveyor belt and then pass through a magnetic system that selects out the ferrous materials, which comprise about five per cent of the total waste.

The ferrous metals are run through a machine that shapes them into small nuggets that are sold for $20 a ton to a southern Illinois steel company for use in place of iron ore in blast furnaces. There are 150 pounds of recoverable steel in every ton of solid waste.

The nonferrous materials, such as aluminum, copper and glass, are trucked to a small landfill on the shredding facility site. The lighter materials which do not drop through the air jet are blown upward into a storage bin. Odorless, colorless and looking very much like confetti, they are loaded onto trucks and hauled 18 miles to Union Electric's Meramec plant in St. Louis County. There they are fed by a pneumatic system into the 140-megawatt boiler along with pulverized coal to generate electricity. According to Union Electric engineer David Klumb, the Energy Recovery Project is the only trash-to-electricity project in the nation that converts garbage to fuel using this shredding process.

Alan Molitor, city project engineer with the Energy Recovery Project, sees the trash-to-energy system as economically feasible as well as environmentally acceptable. "The concept behind the project is basically so very simple that it's natural to ask why someone hasn't implemented it before," Molitor said. "The environmental movement and the increasing national energy demand have all helped to make trash-to-energy systems an idea whose time has come," he continued. "Coal has tripled in price, and we've realized that materials we were throwing away have become valuable."

Union Electric's plan must be approved by the EPA, but Dougherty anticipates easy approval by that agency. Arsen Darnay, EPA assistant deputy administrator for solid waste management programs, has consistently urged cities to consider using solid waste as a fuel to generate electricity. Darnay said the nation could save 150 million barrels of oil a year if 70 to 80 percent of large metropolitan areas' solid waste was burned in systems like Union Electric's.

Dougherty said that the company does not expect opposition to the project from local environmental groups. But Ben Senturia, executive vice-president of the St. Louis Coalition for the Environment, said there are a "number of unanswered questions" about the project and added that he does not view trash-to-energy systems as a "long-range solution" to solid waste disposal problems. "The trash-to-energy project is essentially an incinerator and incinerators have created serious air pollution problems throughout the country," Senturia said. "Testing of the system to determine what air pollutant concentrations exist has been delayed. The system is presently being touted around the country with no available data on what it is putting into the air."

Dougherty said that "no major pollutants" result from burning trash in the system. The sulfur content of the burnable solid waste is about one tenth of one per cent, which is equivalent to the sulfur content of low sulfur fuel oil, according to Union Electric engineers. "The only possible added pollutant may be some increased particulates, but tests have indicated that their effect is minimal," Dougherty said. EPA and Union Electric have been evaluating the prototype's environmental impact, and results of their testing are expected to be available soon.

Senturia also criticized the project for "still representing the 'burn or bury' approach to solid waste disposal" because it does not recycle paper, ceramics, glass and other resources from the trash. But Molitor anticipates that as soon as the technology for recovering these nonferrous materials from solid waste is perfected, such a "total recovery system" could easily be grafted onto the Union Electric system. The system depends heavily on the 40 to 50 percent of the solid waste's total heating value that comes from its paper component, which makes paper an unlikely candidate for recovering and recycling. Dougherty is optimistic that Union Electric will find a ready market for all the recycled materials the expanded system can recover. St. Louis Mayor John Poelker and St. Louis County Supervisor Lawrence Roos have both expressed "enthusiastic support" for the Union Electric project. Poelker said he sees the project as a "wave of the future" and a prototype for trash-to-energy systems that will be copied by other American cities. "A system utilizing trash to generate electricity seems to be the way to go," he said.

5

WHAT POLLUTES OUR AIR? WHERE DOES IT COME FROM? HOW HARMFUL IS IT?

ROBERT L. WOLKE

Everybody talks about air pollution. Everybody has his favorite villain. But nobody seems to know exactly how harmful it is. In Article 6 you will read an authoritative discussion of the current air pollution problem in the United States. But first, by way of orientation, here is a simple graphic illustration of the types of pollutants in our air, their sources, and their relative harmfulness to humans.

In the left-hand portion of Figure 1 (labelled "unadjusted") are shown the major *components* of air pollution, by percentage: NO_x = nitrogen oxides, HC = hydrocarbons, SO_2 = sulfur dioxide, SP = suspended particulates, and CO = carbon monoxide. In the right-hand portion (labelled "adjusted"), these percentages have been multiplied by their relative toxicities; a larger block, therefore, would indicate a substance exerting a greater overall toxic effect, even though its percentage in the air may be relatively small.

In the left-hand portion of Figure 2 (labelled "unadjusted") are shown the *sources* of air pollutants, by percentage. In the right-hand portion (labelled "adjusted"), these percentages have been multiplied by their relative toxicities; the largest block, therefore, represents the source whose products exert the most harmful effects, and so on.

Note that Figure 1 shows that CO is not a very harmful pollutant, relatively speaking, but that SO_2 is about twice as serious as one might guess from its percentage in the air. Particulates are about three times as serious as their abundance alone would indicate. Figure 2 shows that the burning of fuels in stationary sources (such as buildings, power plants, and factories) is a much more significant hazard than transportation sources (such as automobiles and airplanes).

The "snake" diagram (Figure 3) is self-explanatory: it shows how all our energy in the United States is produced and what it is used for. A few minutes' study of this diagram will help put the preceding articles on energy and the following one on air pollution into perspective.

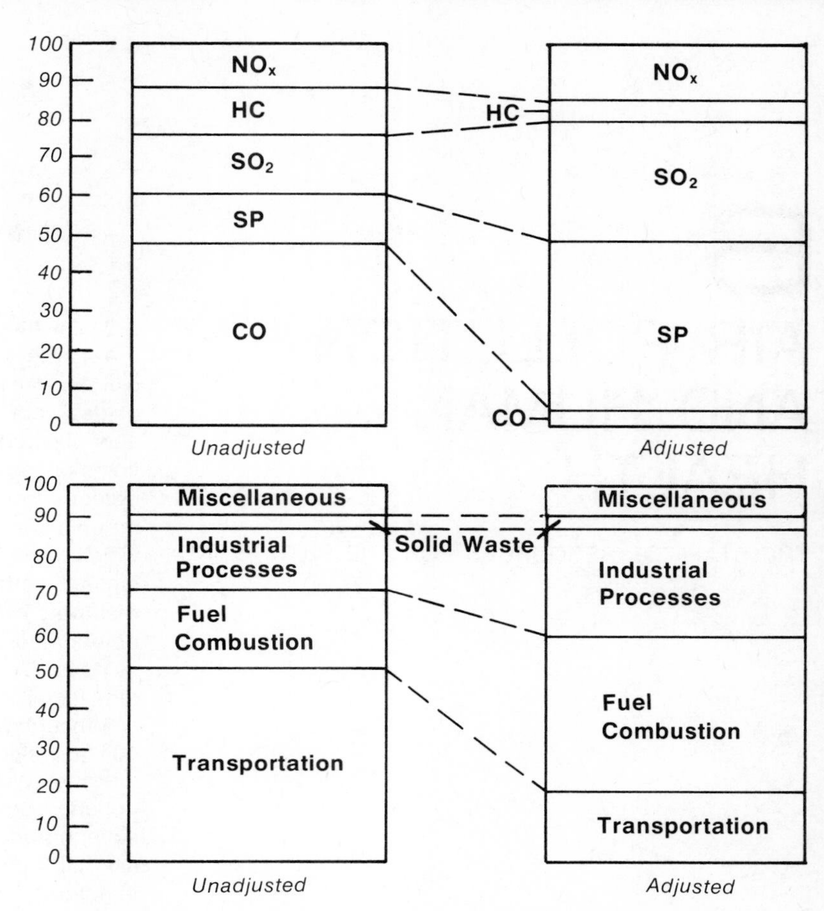

Figure 1. 1971 air pollution emissions, percentage by pollutant, unadjusted and adjusted for effects. (Based on a method developed by Professor Lyndon Babcock of the University of Illinois; adapted from "Environmental Quality 1973—The Fourth Annual Report of the Council on Environmental Quality," Washington, D.C.)

Figure 2. 1971 air pollution emissions, percentage by source, unadjusted and adjusted for effects. (Based on a method developed by Professor Lyndon Babcock of the University of Illinois; adapted from "Environmental Quality 1973—The Fourth Annual Report of the Council on Environmental Quality," Washington, D.C.)

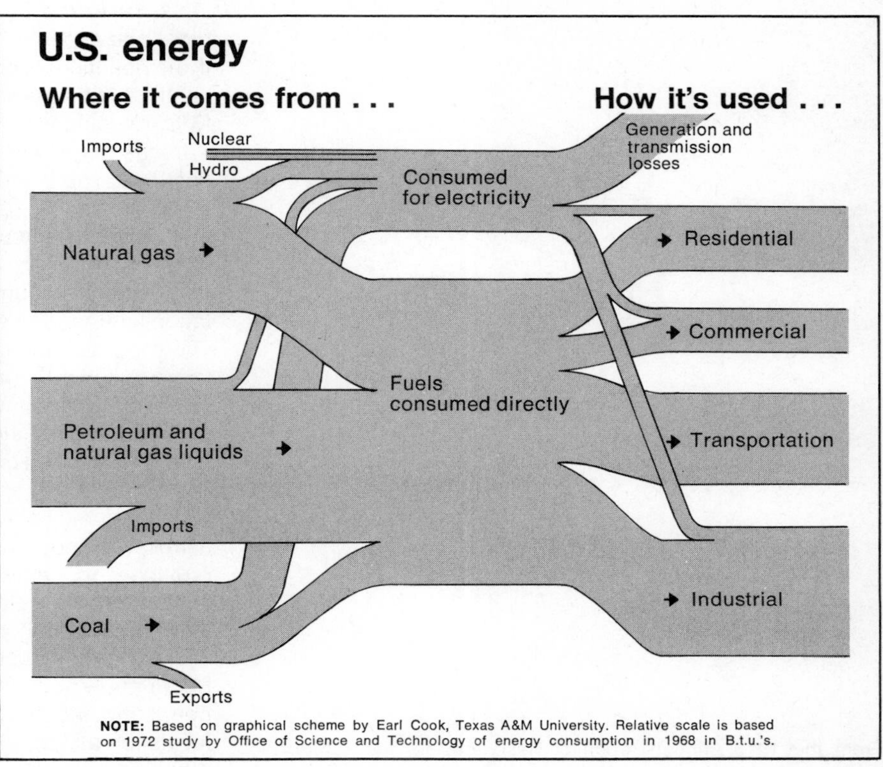

Figure 3. The sources and uses of energy in the United States. (From *Chemical and Engineering News,* November 13, 1972. Reprinted with permission of the copyright owner, the American Chemical Society.)

6
AIR POLLUTION AND HUMAN HEALTH

THE STAFF OF RESOURCES FOR THE FUTURE, INC.

From the 1973 Annual Report of Resources for the Future, Inc. Adapted from a report by Lester B. Lave and Eugene P. Seskin.

It is misleading to speak of air pollution as if a single substance were involved. Sulfur oxides, suspended particulates, nitrogen oxides, carbon monoxide, hydrocarbons, and photochemical oxidants are the most common pollutants in the air. Lumping together all of these into the single category, air pollution, provides little information as to which specific pollutants should be abated, since the effects include phenomena of such disparate weight as discoloration of fabrics, deterioration of health, and fewer clear, sunny days. Air pollution is used as a summary term for these compound phenomena.

Although many of the effects of air pollution are evident, they are often difficult to measure. We do not know precisely what air pollution costs society in terms of human suffering and deprivation, not to mention financial injury. Many investigators have attempted to quantify and evaluate these effects. Lave and Seskin have elected to explore one aspect in detail, conjecturing that, if air pollution exerts an adverse effect on health, this alone would be sufficient to justify its abatement.

Let us begin with some useful oversimplifications. There are two types of air pollutants—those that come mostly from stationary sources (particulates, sulfur dioxide, and nitrogen oxides), and those that come mostly from mobile sources (carbon monoxide, lead, hydrocarbons, and nitrogen oxides). There are two measures of ill health—sickness and death (morbidity and mortality). And there are two ways by which the health problems are presumed to arise from air pollution—chronic exposure and episodic exposure.

In living memory, several large-scale, disastrous episodes have provided the kind of public evidence needed to focus official attention on the relationship between daily mortality and daily air pollution.

The highly industrialized Meuse valley of Belgium experienced climatic conditions permitting the buildup of abnormally high levels of air pollutants (particularly sulfur dioxide) during December 1930. Over a five-day period, approximately 6,000 people became ill and perhaps 60 died (most of whom were elderly persons or those with previous heart and lung conditions). This was more than ten times the number of deaths that would normally be expected.

In October 1948, a similar event took place in Donora, Pennsylvania. Within three days, almost 6,000 people (over 40 percent of the population) became ill, and about twenty deaths were reported. This again was approximately ten times the expected number of deaths, and again the aged were most susceptible (the average age of the dead was 65). A follow-up study found that persons who became ill during the smog episode demonstrated subsequently highr mortality and morbidity than other persons living in the community. This was particularly true for persons with diseases of the cardiovascular or respiratory systems.

In December 1952, London was enveloped by a dense fog and in a two-week period 4,000 deaths were attributed to the abnormally high concentrations of sulfur dioxide and smoke. Unlike the previous epsiodes, all age groups were affected.

Lesser episodes occurred in London during January 1956 and December 1962. Nearly 1,000 deaths were attributed to the first incident. The greatest percentage increase in deaths was among newborn children. In absolute numbers, however, the greatest increase was among elderly persons. Bronchitis mortality exhibited the highest increase among specific causes of death.

Dramatic increases in air pollution have been shown beyond question to cause discomfort, illness, and death. Thus, the first hypothesized relationship between air pollution and health is an acute response, in which high concentrations of air pollutants have an immediate effect on health. Since there are a host of pollutants which might be responsible, some researchers have examined air pollution episodes to determine which pollutants were present, and numerous laboratory experiments have been run to explore the effects of individual pollutants.

The second hypothesized relationship is a more subtle one . . . The effect of air pollution on health may be related to exposure time. A long, or chronic, exposure to low concentrations might be just as harmful to health as a short, or episodic, exposure to high concentrations. This long-term exposure may well exacerbate existing disease or increase susceptibility to disease. Evidence for this hypothesis is the finding that such substances as benzopyrene can produce cancer in laboratory animals, and the fact that acute irritation (demonstrated in laboratory experiments) can aggravate the symptoms of a chronic respiratory disease, and possibly make it progressive. However, concentrations of the magnitude required to demonstrate these effects in laboratories are seldom, if ever, experienced in urban air.

While there is no doubt that air pollution did cause illness and death in the episodes of Donora and London, there has been considerable uncertainty as to whether the levels of air pollution now prevailing in our major cities are sufficient to affect health. Although people living in cities are known to be generally less healthy (with lower life expectancies) than people living in rural areas, the question remains whether any part of this "urban factor" is caused by air pollution. Detailed and statistically sophisticated analyses are needed to explore the relationship between the level of air pollution and number of deaths. This research must not only establish whether there is an effect, but must also provide estimates of the quantitative impact of air pollution on the health of city populations.

We need to know, specifically, whether the effect of a given dosage changes with the time of exposure. For example, is an exposure of 10 parts per million (ppm) of sulfur dioxide for 10 hours better or worse than an exposure of 1 ppm for 100 hours or 0.1 ppm for 1,000 hours? If the physical responses to these exposures are quite different, laboratory experiments are not likely to provide satisfactory information for public policy toward air pollution. The body's repair mechanism has led to the prevalent view that there is a "threshold" above which the repair mechanism can no longer keep up with the damage; below the threshold level, no damage will result. If there exists a threshold level of air pollution that can be easily maintained by society, it makes sense for Congress to pass laws requiring that air pollution be abated to the point where no health damage is present. Such an attempt was indeed made in the 1970 amendments to the Clean Air Act. But if no such threshold exists, or if it is at so low a level that it cannot be maintained with current technology, the stan-

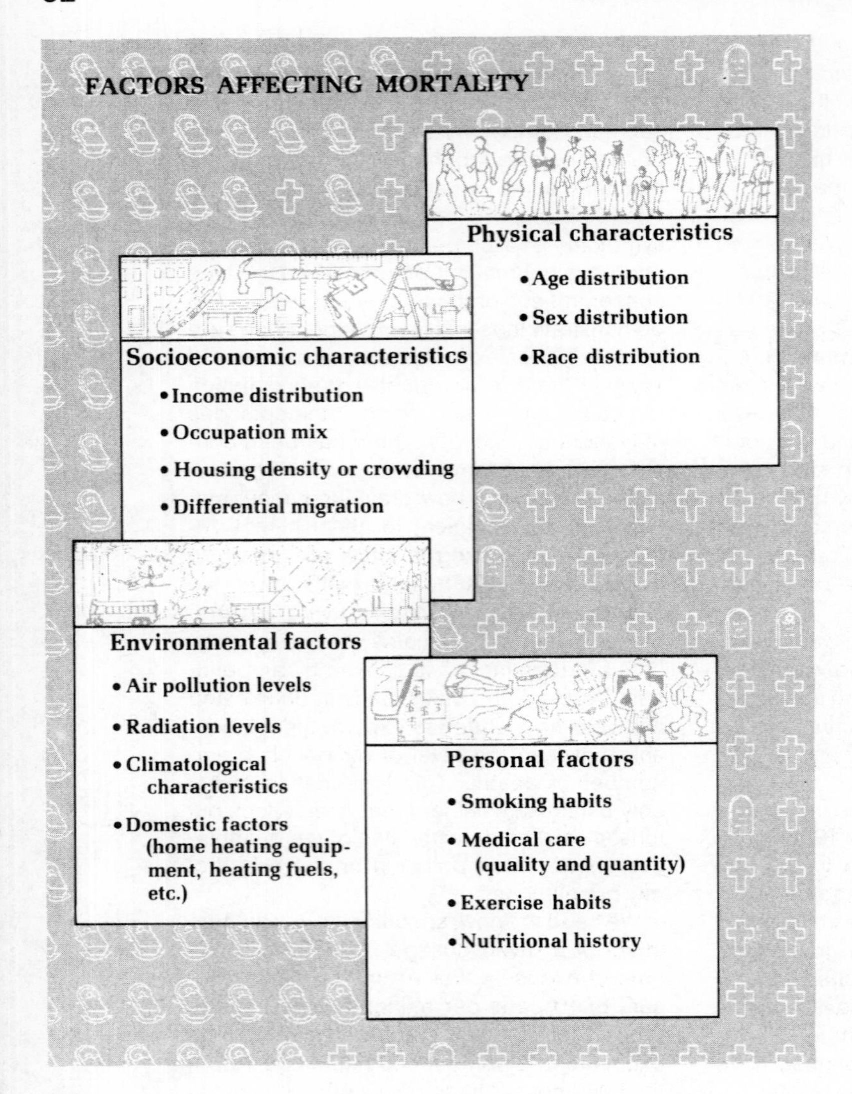

dards embodied in the 1970 amendments make little sense.

Though not yet agreed on such a threshold, scientists have been accumulating formal evidence for more than forty years that links ill health to air pollution. The scientific community has been slow to accept this evidence because of the methods used in gathering it and because of the lack of controls. Some early studies contrasted the mortality rates in polluted and unpolluted areas and found higher death rates in the polluted regions. The problem with such analyses is that areas with high levels of air pollution are industrialized cities, whereas the areas with low levels of air pollution are rural farming communities. Put this way, the findings are not at all surprising: we know that people living in large cities have lower life expectancies for a host of reasons.

The principal conclusion that one might draw from a literature review is that many scientists have demonstrated an association or link between air pollution and increased mortality rates. What can one conclude from such a link? There are four logical possibilities: (1) the association is a sampling phenomenon and occurs at random; (2) air pollution causes an increase in the mortality rate; (3) increases in the mortality rate cause air pollution; or (4) there is a third factor that causes both air pollution and increased mortality, which gives rise to a spurious correlation between them. For example, if automobiles were the only source of air pollution, and if many deaths resulted from automobile accidents, there would be a significant correlation between air pollution and the mortality rate across cities. To avoid this spurious association, the number of automobiles, as well as such related factors as the number of miles driven and weather conditions, would have to be controlled in the analysis.

What can we say about these four possibilities? We can rule out the first, since there has been an enormous volume of evidence collected indicating a close association between air pollution and increased mortality. We conjecture that the second possibility is correct, but we must rule out the third and fourth in order to prove it. The third possibility seems macabre: perhaps in the 13th century the increased mortality during the Black Palgue epidemics led to air pollution, but this hardly seems relevant today. Therefore, proving that air pollution is the cause of increased mortality depends upon being able to rule out the fourth possibility, i.e., that the association is spurious.

Generally, mortality rate is measured for a geographically defined group, such as the inhabitants of a city. A large number of factors are known to affect the mortality rate. Some of these may be grouped arbitrarily into: (a) physical characteristics of the population, (b) socioeconomic characteristics, (c) environmental factors, and (d) personal factors. The most important factors are the last, yet they are the most difficult for researchers to take into account.

In order to estimate the effect of any one of these factors on the mortality rate, the others must be held constant experimentally or controlled statistically. This is necessary because the estimated difference in mortality between two areas may reflect differences in a variety of factors affecting the death rate; one cannot ascribe the entire difference to one, such as air pollution.

An ideal investigation of the association between air pollution and health would control for all relevant factors. Unfortunately, many of them are difficult to measure conceptually (genetic effects, for example), while data on others either have not been collected (smoking habits) or are poorly measured in existing statistics (medical care). The difficulty comes in finding ways to control for each of the variables experimentally or statistically, explicitly or implicitly.

In practice, the laboratory has not proved to be a very useful setting for investigating the association between air pollution and ill health. There need be little relationship between the short-term reaction mechanisms which can be explored in the laboratory using human subjects, and the mechanisms by which exposure to low levels of pollution over a period of many years exacerbates chronic disease. Even experiments involving the fumigation of animals over a prolonged period may have little relevance in studying the effects on humans, because of the differences in physiology and life span.

Because of these difficulties, the normally accepted method of using laboratory experimentation to prove causality may not be suitable. Indeed, the notion of causality may not be a useful concept in investigating chronic disease. Progressive lung damage could be caused by bacteria (or a virus), by occupational exposure to harmful materials, by smoking, by air pollution, or even by genetic factors. What is the cause of severe dyspnea in a 60-year-old asbestos worker who smokes, lives in a large city, comes from an impoverished family, and has never received proper nutrition?

A more useful approach for chronic disease involves determining which factors aggravate symptoms or make the disease progressive. In the given example, all of the factors named probably contributed to the dyspnea, and the absence of any one of them probably would have resulted in less severe symptoms. But the worker would not find such a conclusion helpful, nor would a public health official. They would find it more useful to know, for example, that breathing asbestos particles is ten times more detrimental than exposure to urban air pollution at the level experienced in the particular city.

Similarly, it is more useful to determine the extent to which sulfur dioxide increases the severity and frequency of emphysema than to investigate whether high concentrations of sulfur dioxide induce the disease in white mice under controlled conditions. Lave and Seskin's concern is primarily with estimating the effect of mitigating one of the insults, rather than with whether this insult is capable of causing the disease in isolation. For this reason, an epidemiological study that examines human beings in their natural setting is more relevant than a laboratory experiment. But to investigate the contribution of one insult, one must control for all the other variables believed to influence health. And research is further complicated by the likelihood that pollutants interact to produce more damage than each would individually. The effects of uncontrolled or unobserved variables will sometimes result in lending a spurious significance to the air pollution—health relationship; in other cases, in obscuring the association.

For a chronic disease, such as bronchitis, there is a long series of possible causes and aggravating factors. It is difficult, if not impossible, to control for each of them, e.g., genetic susceptibility. One is faced with a choice between two types of experiments.

The first type of experiment respresents classical epidemiology; it compares two groups which differ only in their exposure to air pollution. For example, one might look at the incidence of bronchitis among sets of identical twins, one of each set living in an area of low air pollution and the other living in an area of high air pollution. Studies of this sort may be quite suggestive, but they are not very helpful in estimating the contribution of air pollution to the incidence of bronchitis in the general population. These experiments also suffer from small sample sizes and limited ability to control for other factors hypothesized to affect the incidence of ill health; for example, a difference in exercise habits between urban and rural environments.

The second type of study examines the incidence of bronchitis among large, geographically defined groups, and searches for correlations among a variety of possible causes. Sample size is particularly important when a subtle irritant is involved and a number of other variables are uncontrolled. For example, in comparisons of the populations of areas exposed to different levels of air pollution, the inherent health-related factors will tend to balance out. When this is the case, a comprison between two substantial populations can be expected to show different mortality, morbidity, or physical-test results, if air pollution is the prime variable.

Of course, data on the characteristics of various city populations may not reveal some crucial traits needed for valid comparisons. People with severe respiratory difficulties may migrate from polluted to unpolluted cities. Suppose, for example, that everyone in New York City who has severe asthma or another chronic respiratory disease moves to an unpolluted city in the Southwest, such as Phoenix. This would mean that, even if the relatively clean air in Phoenix were to prolong the migrants' lives, the net effect of the migration of unhealthy people might be to raise the mortality rate in Phoenix while lowering the mortality rate in New York City. If there is a net migration of diseased people toward unpolluted areas, . . . [this will] cause us to underestimate the association between air pollution and mortality. Carried to the extreme, so many sick people could migrate to Phoenix that the resulting death rates would be higher in Phoenix than in New York, and the estimated relationship between mortality and air pollution would show that air pollution is associated with lower mortality! The effects of migration have in fact been explored by Lave and Seskin in an analysis of the change in population in standard metropolitan statistical areas (SMSAs) between 1950 and 1960. Migration appeared to have little bearing on the estimated effects of air pollution.

A similar distortion might arguably occur when people take less dramatic steps to reduce their exposure to air pollution. If persons with respiratory diseases move within a city to a less polluted area, or if they install devices in their present homes to filter out pollution, the mortality rate is likely to fall. Behavior of this sort will affect the estimated association with air pollution. Insofar as such actions take place with greater frequency in more polluted cities, the mortality rate in these cities will be lower than "expected."

Another serious issue involves the error associated with air pollution measurements. In general, a single sampling station in a region generates readings that are reported biweekly. Since pollution concentration varies greatly with terrain, it is an heroic assumption to regard these figures as representative of an entire SMSA in making comparisons between metropolitan areas. Furthermore, measuring instruments change over time and differ from city to city, with some instruments having little reliability. At best, these air pollution reports are a remote approximation of the true current and cumulative exposure to air pollution experienced by the inhabitants.

Notwithstanding conceptual and technical problems of this magnitude, sound research has been carried out in the United States and elsewhere . . . The vast bulk of this [research] reports a significant association between air pollution and health. The recently completed research by Lave and Seskin gives much added weight and authority to that general conclusion.

Two statistical approaches were adopted, one involving broad-scale use of data on a great many cities, the other focusing narrowly on a handful of cities. In reviewing the broad-scale work, it should be conceded that the data available on particular cities could be seriously at fault. Yet, overall, it is believed that these deficiencies balance out and permit valid generalizations to be drawn for American cities taken as a whole.

For the year 1960 the mortality rates of 117 SMSAs were compiled so that the effect of air pollution could be estimated; socio-economic variables were used to control for population density, racial composition, age distribution, and income. The two pollutants under consideration were suspended particulates and sulfates. The conclusion reached was that air pollution was an important factor in explaining variation in the total death rate across areas of the United States. To corroborate these findings, a partial replication was performed, using 1961 mortality and suspended particulate data, which gave largely similar results.

The analysis was then extended by disaggregating the total death rate into infant mortality, age, sex, race-specific mortality, and disease-specific mortality. A number of important problems that tended to obfuscate the association between air pollution and these disaggregated mortality rates had to be dealt with. These included possible underreporting of the fetal mortality rate, inaccuracies in the diagnosis for the cause of death in disease-specific rates, and sampling variability when the disease was relatively rare (and the mortality rate was small) or when the population at risk was not very large.

Among the infant rates, particulate pollution was most closely associated with the deaths of infants under one year and under 28 days old, while sulfate pollution was most closely associated with the fetal death rate. In all, twenty-nine age, sex, race-specific mortality rates averaged over the years 1959 to 1961 were investigated. In general (except for the 15- to 44-year-old age groups), air pollution was an important factor in explaining the variation in these mortality rates across SMSAs.

"You win a little and you lose a little. Yesterday the air didn't look as good, but it smelled better."

Sidney Harris, American Scientist.

Fourteen disease-specific mortality rates, based on both 1960 and 1961 data, were examined. For total cancers and four subclassifications of cancers, there was a close association with air pollution (particularly sulfates) in 1960, although the 1961 replication failed to indicate the same level of significance. Deaths from cardiovascular disease and its subcategories also showed a close association with sulfate pollution in 1960; in this case, the 1961 results further strengthened the relationship. For five diseases of the respiratory system, the results were generally insignificant. Only the tuberculosis mortality rate exhibited a close association with air pollution in both 1960 and 1961. Surprisingly, the mortality rates for asthma, influenza, pneumonia, and bronchitis did not display a relationship that was significant in both years.

For a subsample of fifteen metropolitan areas, data on three additional pollutants (sulfur dioxide, nitrates, and nitrogen dioxide) were investigated in an attempt to isolate their effects. In addition, major interactions among the pollutants were analyzed. Sulfates proved to be the deadliest pollutant in terms of general mortality. However, for the infant mortality rate, all five pollutants were statistically significant, and nitrogen dioxide was the most important in terms of estimated effect.

In work focusing more narrowly on a small group of major cities, greater specificity was gained but generalizing power was much reduced. As with all previous work on this problem, this study was limited by the availability of pollution data and, to a lesser extent, of mortality data. Daily observations on pollution levels (24-hour averages) were obtained from the Continuous Air Monitoring Program (CAMP). The data were most usable and complete for five cities—Chicago, Denver, Philadelphia, St. Louis, and Washington, D.C.—and for five pollutants—carbon monoxide, nitric oxide, nitrogen dioxide, sulfur dioxide, and hydrocarbons.

In conjunction with the daily pollution readings, daily death counts from 1962 to 1966 were obtained for the five cities from a special study by the Environmental Protection Agency.

Finally, climatological data were secured on daily weather factors for the cities from the Department of Commerce. Air pollution, mortality, and climate information constituted the data base.

Results were similar for four of the five cities. There was no evidence of an effect of daily air pollution on daily mortality. The notable exception was Chicago, where, after controlling for climate and weekly effects, a close association was found between daily deaths and two pollutants—nitric oxide (primarily from automobiles) and sulfur dioxide (primarily from stationary sources). The effects of these pollutants on Chicago's population differed in their severity and timing. Sulfur dioxide was the most deadly, showing a large immediate effect on mortality, which gradually diminished over time. For nitric oxide the effect was delayed by several days, reached a peak, and then diminished. Further analysis ruled out the likelihood that deaths associated with air pollution were merely hastened by a few days—that is, that very vulnerable or mortally ill individuals who would have died a bit later were simply prodded along by the pollutants.

Some climate variables were significant in each city. Average wind speed was significant only for St. Louis, where increased wind velocity seemed to reduce daily mortality. This was plausible, since wind cleanses the air. Rainfall did not appear to have this effect. However, a lower mean temperature (coupled with air pollution) produced a rise in the death rate in Denver and St. Louis, whereas a higher mean temperature produced more deaths in Chicago. It is difficult to explain the different patterns in the three cities. For Chicago, however, it is possible to say that a 50 percent reduction in air pollution (as measured by sulfur dioxide) could be associated with a 5.4 percent reduction in that city's daily death rate.

A natural question arises as to why the data from Chicago alone confirmed a connection between daily mortality and daily pollution levels. One possibility concerns the relatively high levels of pollution that were prevalent in Chicago. For each pollutant under consideration, the mean level in Chicago exceeded the mean value for any other city. This was especially noteworthy for sulfur dioxide, for which Chicago's mean levels were almost ten times the levels of Denver and almost four times the levels of St. Louis. Even the low pollution months in Chicago had pollution levels far greater than those of the other four cities in this study. Thus, it may be that, at levels of pollution substantially below those found in Chicago, acute effects are not important. A closely related issue involves the relative size of the cities. Because of its larger population, the mean number of deaths per day in Chicago was more than four times as large as in any other city in the sample. Since deaths occur in discrete units (at least one at a time), effects of pollution in a smaller city may be lost when scattered over a five-day period.

Other studies, it should be added, have found similar close associations between daily mortality and air pollution in New York City and Los Angeles, both of which answer the requirements of size and high levels of air pollution.

Before Lave and Seskin undertook the foregoing analysis, they examined a large number of studies documenting an association between air pollution and ill health. The conclusion from both their work and the work of others is that a significant association exists between various pollutants and death rates in large cities across the United States. Moreover, it appears that daily changes in the cities' air pollution levels have more effect on mortality than do annual changes. Where possible, every reasonable suggestion was tried to determine if the estimated relation between air pollution and mortality was spurious. In general, the conclusion has passed these tests and appears to be a consistent one. While this evidence cannot be said to prove causation, since there are still a number of factors affecting health which were uncontrolled in the analysis, it is substantial and, it is believed, sufficient to convince an unbiased observer of the relationship.

The implication of the 1960 study for 117 SMSAs is that a 10 percent reduction in air pollution (as measured by particulates and sulfates) was associated with a 0.90 percent decrease in the total mortality rate. Analysis of 1960 data also suggests that a 50 percent reduction in air pollution (as measured by particulates and sulfates) would be associated with a 4.5 percent decrease in the total mortality rate. When a similar calculation was made from the corresponding 1969 data, it was found that a 50 percent reduction in air pollution (as measured by particulates and sulfur dioxide) would be associated with a 7.0 percent decrease in the total death rate.

This work points to the need for a great deal of additional research to answer specific questions. An investigation of morbidity rates would probably be more useful than further studies of mortality rates. It is essential that the more difficult task of constructing reliable morbidity indices be accomplished. . . . Another crucial issue involves the question of *which* pollutants are harmful to health. For example, what is the toxicity of photochemical smog compared with that of sulfur dioxide?

Some limited evidence has been produced that suggests that suspended particulates and sulfur oxides are important pollutants. For infants, suggestive although quite limited results lead to the conclusion that nitrogen oxides as well as particulates are quite toxic. Much more work must be done on toxicity, and it is likely that such investigations will require morbidity rather than mortality data.

Finally, the problem of estimating the effect of air pollution on human health requires better physiological evidence and identification of other environmental insults, such as radiation. A program should be undertaken immediately to estimate the doses of environmental factors experienced by a sample of the population, and to correlate these estimates with measures of the morbidity of the same sample. A careful experimental design together with appropriate statistical analysis of the data would add much to our understanding of the effects of man's environment on man.

7 FUTURE USE OF MINERALS: THE QUESTION OF "DEMAND"

JOHN D. MORGAN, JR.

The earth and the sun are the fundamental sources of all wealth. Demand is limited absolutely by what can be obtained from the rocks, soils, waters, and atmosphere of the earth, plus earth's share of solar energy. Rising standards of living in the technologically advanced nations, coupled with the desires of the burgeoning populations of the less developed nations, focus increased attention on demand for minerals, both now and into the foreseeable future.

Man's technological progress has been so dependent upon minerals that major historical periods are named the Stone Age, the Bronze Age, the Iron Age, the Steel Age, the Age of Light Metals, and the Nuclear Age, marking the new mineral materials added to those that had been in use from the dim past of prehistory. However, in terms of tonnages in use and annual new tonnages required, we are still in the Stone Age, because about half of United States annual demand for new mineral supplies involves sand, gravel, and stone.

The original inhabitants of the three and a half million square miles that today is the United States used a number of minerals, including clay, natural asphalts, rock oil, red and yellow iron oxides, granite, flint, and turquoise. But their total annual demand for all those minerals would have totaled only a few tens of tons. By 1950 United States annual demand for new mineral supplies including fuels had reached two billion tons; by 1973 it had doubled to four billion tons; and projections of demand point to annual United States demand of eleven billion tons by the year 2000.

It is important to consider demand not just in overall quantitative terms but also in value terms . . . Figure 1 shows the role that minerals play in the United States economy. . . . While it becomes increasingly difficult to identify the values attributable to minerals *per se* as the stages of manufacturing become more advanced, it is obvious that minerals and agricultural products together are the fundamental materials that make possi-

From "The Mineral Position of the United States, 1975–2000", Eugene N. Cameron, Ed. Published for the Society of Economic Geologists Foundation, Inc., by the University of Wisconsin Press, Madison, Wisconsin, 1973. (Original data updated by the author).

THE ROLE OF MINERALS IN THE U.S. ECONOMY

(ESTIMATED VALUES FOR 1973)

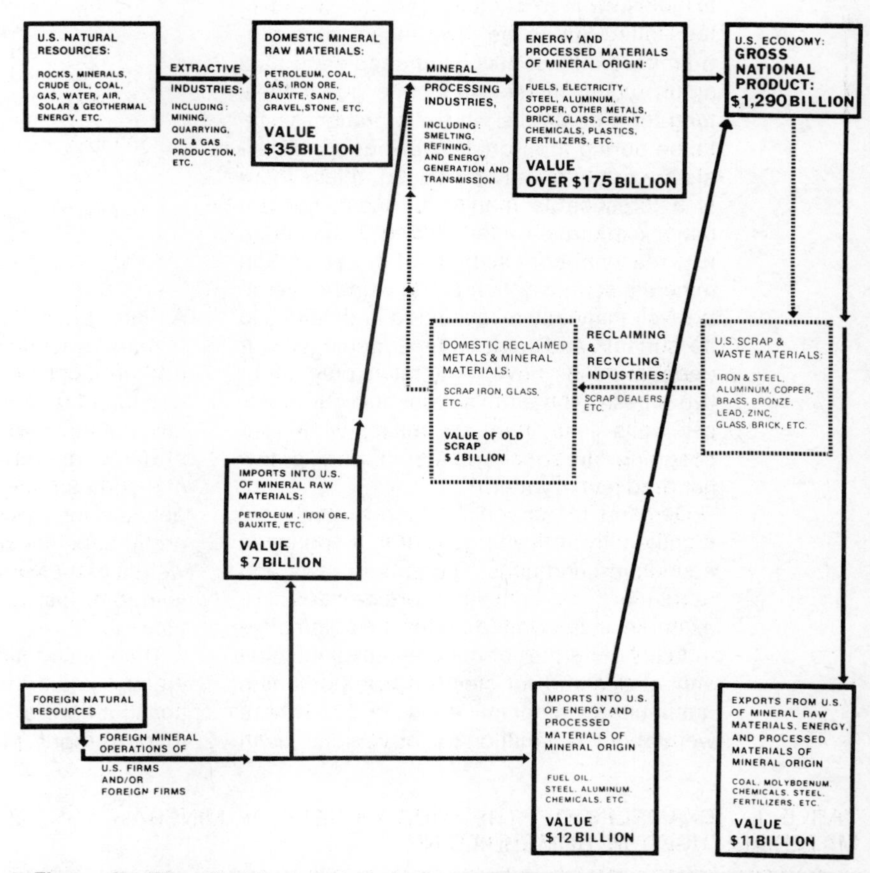

Figure 1. The role of minerals in the United States economy (estimated values for 1973).

ble the United States Gross National Product, which was valued at $1,290 billion in 1973.

National security considerations must be broadly viewed. Politico-economic factors must be weighed along with purely military ones. The United States should not allow itself to become too dependent upon potentially unreliable sources of supplies. Remedial measures could include maintenance of some assured level of domestic production, stand-by facilities, stockpiles, development of substitutes, and/or standby controls. National security considerations, broadly viewed, may also place special politico-economic restrictions on mineral demand; for example, favoring demands for minerals produced in some sources while discouraging demands upon other sources. Balance-of-payments considerations can affect demand for minerals which contribute either positively or negatively thereto. Politico-economic considerations affect demand for monetary metals. . . .

Real demand for minerals must bear some relationship to actual ability to pay for them and the ability of the earth to supply them. Further, real demand and real supply are by no means likely to be in phase. Indeed, except in defense emergencies, over past decades and continuing to the present, short-term availabilities have been such that producers of most minerals have made major efforts in research, advertising, and marketing to increase demand in old and in new applications. In some instances, most often in defense emergencies, national demands for defense programs were quantified by governments, and major supply-expansion programs were initiated to increase supplies to meet new levels of anticipated demand. In the United States the most recent such programs were those involving accelerated tax amortization, market and price guarantees, long-term contracts, and exploration assistance during the Korean War period. Minerals are not ordinarily produced unless there is a foreseeable market for them, nor are major exploration efforts made unless there is a reasonable possibility of marketing the minerals sought. Much of the world is yet to be systematically investigated in detail as to its surface and near-surface geology. Our deepest mines have penetrated only about two miles of the earth's crust and our deepest wells less than six miles, while our seagoing dredges operate in only a few hundred feet of water.

Demand for specific minerals may vary significantly from year to year in response to economic conditions, changes in laws and regulations, or consumer preferences. For example, in the United States the automotive industry generates about one-fifth of the total annual demand for steel and proportionate quantities of other minerals. In 1973 there were about 120 million motor vehicles of all types on the nation's roadways. While the specifications for vehicles vary widely, the average standard American auto is made up approximately as follows:

2,775 lbs.	iron and steel (frame, engine, and body)
100 lbs.	aluminum (components and some engines)
50 lbs.	copper (radiators and electrical)
25 lbs.	lead (battery)
50 lbs.	zinc (castings, galvanizing, and tires)
100 lbs.	glass (windshield and windows)
250 lbs.	rubber and plastics (tires and fittings)
150 lbs.	miscellaneous (fabric, insulation, etc.)
3,500 lbs.	Total

Almost all of the above are derived from minerals, including the plastic and synthetic rubber. Consequently, American 1973 production of 9.7 million autos, plus 3.0 million new commercial vehicles, generated mineral demands substantially greater than 12.7 million times the above unit quantities. The demand for processed materials of mineral origin substantially exceeds the drive-away weight of the vehicle because there are unavoidable losses in manufacturing operations.

The construction industry is another generator of major demands for minerals. When construction of houses, hotels, factories, roads, airports, power plants, harbors, etc., is

TABLE 1. [EXAMPLES OF] THE WIDE VARIETY OF MINERAL MATERIALS USED IN HOMEBUILDING

Building materials	Insulating materials	Heating and cooling materials	Plumbing materials
Sandstone	Asbestos	Burning coal (heat only)	Plastic pipe
Limestone	Rock wool	Burning oil	Soft copper tubing
Marble	Fiberglass	Burning natural gas	Copper pipe
Granite	Plaster (gypsum)	Burning manufactured gas	Brass pipe
Concrete	Plaster board (gypsum board)	Electricity, in turn derived from:	Black iron pipe
Common brick	Vermiculite	Coal	Galvanized iron pipe (zinc-coated)
Tile	Fly-ash block	Oil	Cast iron pipe
Concrete block	Aluminum foil	Natural gas	Lead pipe (waste lines only)
Pebbles set in cement	Aluminum shade screening	Manufactured gas	Lead sheet (showers, etc.)
Asbestos-cement siding	Bronze shade screening	Hydropower	Glass pipe
Slate granule siding	Tar paper	Geothermal power	Zinc die-castings
Painted steel	Double pane glass	Nuclear power	Chromium-plated brass
Galvanized steel (zinc-coated)	Plastic sheet	Solar Energy (rarely by itself)	Nickel-plated brass
Oxidizing steel	Foamed plastic		Chromium-plated iron
Stainless steel (chromium-nickel)	Magnesite		Stainless steel
Plastic-coated steel	Foamed glass		Vitreous china
Glass			Porcelain ware
Aluminum			Porcelain-coated steel
Plastic-coated aluminum			Cement-asbestos pipe
Solid plastic			Concrete pipe
Fiberglass plastics			Clay tile pipe
Asphalt shingles			Plastic sheet
Asphalt roofing			Ceramic tile
			Plastic parts and fittings

accelerating, the effect is felt throughout the mineral industry. . . . Table 1 illustrates the wide variety of materials that are used in building ordinary houses and in heating them. All of these materials are of mineral origin. The choice of a given material for a particular use is governed partly by price, partly by specific properties and specifications, and partly by assured availability.

Alternate methods and procedures must also be considered in assessing demand. For example, food may be preserved by packing in "tin" cans, which are actually tin-coated steel. Food may also be preserved by packing in aluminum cans or lacquer-lined steel cans, or in glass jars or bottles. Beverages are packaged in all-aluminum or steel-aluminum cans, or in glass or plastic bottles. Many foods may be preserved by freezing, wrapped in plastic, paper, or aluminum foil. Some foods are still preserved by the older methods of smoking, salting, baking, or drying. Demand in a variety of industries can be achieved by a variety of means.

Export demand is another item that must be considered. The United States in the past has been an important exporter of several mineral raw materials, notably coal, molybdenum, phosphate, sulfur, and vanadium, and many processed materials of mineral origin, notably metals and chemicals.

Not all demands for minerals are increasing. Some indeed may well decline. More questions are being raised as to whether there may not be too much of some items in circulation in certain applications, or perhaps misapplications. In recent years radioactive materials, sulfur oxides, carbon monoxide, nitrogen oxides, hydrocarbons, photochemical oxidants, various particulates in the air, mercury, lead, cadmium, beryllium, phosphorus, asbestos, and many chemicals are among materials subject to restrictions which should curtail demand. Conversely, pressures to improve the environment may stimulate demand for other materials. There is also increasing concern about displaced earth and rock in connection with slides, leaching, and erosion. Any overall calculations of future mineral demand must include realistic balancing of demands for cleaner air, cleaner water, restoration of mined lands, extinguishment of mine fires, rehabilitation of tailing piles, maintenance of wilderness areas, parks, and recreation sites, and disposition of large stocks of materials accumulated pursuant to environmental regulations—as in the case of sulfur.

Steel production is a major index of modern industrialization. Iron, the predecessor of steel and the most important element in steel, has been in use by man for over four thousand years. Iron is one of the more abundant elements, accounting for over 5 percent of the earth's crust. Steel possesses a reasonably high melting point, and its alloys are extremely tough and resistant to most physical forces. While steel manufacturing often involves mammoth installations, its technology is relatively simple, and steel is a relatively cheap material . . . Figure 2 shows the rise of U.S. steel use over the past two decades.

Another major group of metals also fundamental to all industrialized economies consists of the older nonferrous metals copper, lead, and zinc, plus the newer metal, aluminum. Copper has been in use for over five thousand years. Bronze, about 85 percent copper and 15 percent tin, gave its name to an early period. Strong corrosion-resistant bronze alloys of today are little different from those made in Phoenicia four thousand years go. Lead has been in use for over two thousand years, and many current techniques were widely employed in ancient Rome. Zinc has been in use in the Western world for over five hundred years and in China long before. The manufacture and use of white zinc oxide was described by Marco Polo, while the composition of old Chinese brass, about 70 percent copper and 30 percent zinc, is the current formulation for modern brass alloys. Copper, lead, and zinc, however, are relatively rare elements of the earth's crust, all three together totaling less than a small fraction of one percent. In contrast, aluminum, first produced commercially toward the close of the nineteenth century, accounts for over 8 percent of the earth's crust. Aluminum is the third most abundant element in the earth's crust, being exceeded only by silicon (28 percent) and oxygen (47 percent) . . .

In the past two decades, as shown by Figure 3, there has been a very rapid rise in the demand for plastics. Most plastics are mineral-based products, in that they are composed largely of carbon and hydrogen derived primarily from petroleum and natural gas. Fiberglass reinforcing of plastics creates another demand for minerals. The annual tonnage of plastics used in the United States now exceeds the combined tonnages

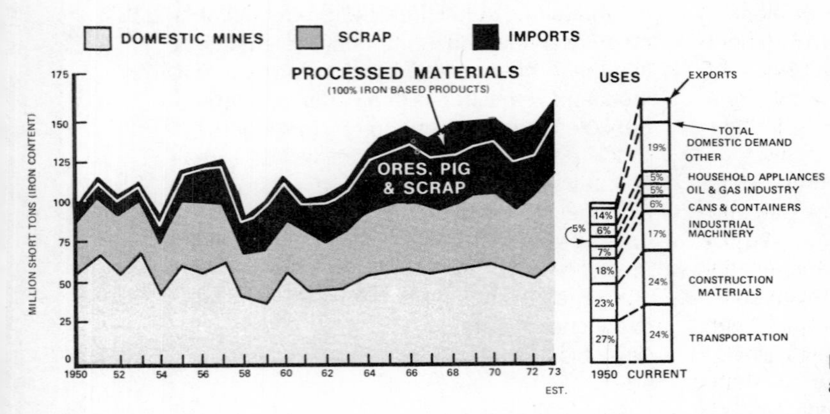

Figure 2. United States supplies and uses of iron, 1950–1973.

of aluminum, copper, lead, and zinc. Inasmuch as most plastics are much less dense than metals, they are used where volume or weight considerations are significant. Because of the differences in density it is possible for a plastic, sold at significantly higher per pound prices, to substitute for a lower priced metal, and on a volume basis United States annual demand for plastics is equal to about half of demand for steel.

Major demands for nonmetallic minerals are generated by the construction and the chemical industries. Our bountiful agriculture is in large measure attributable to abundant use of mineral fertilizers. Demand for refractories and abrasives creates addition-

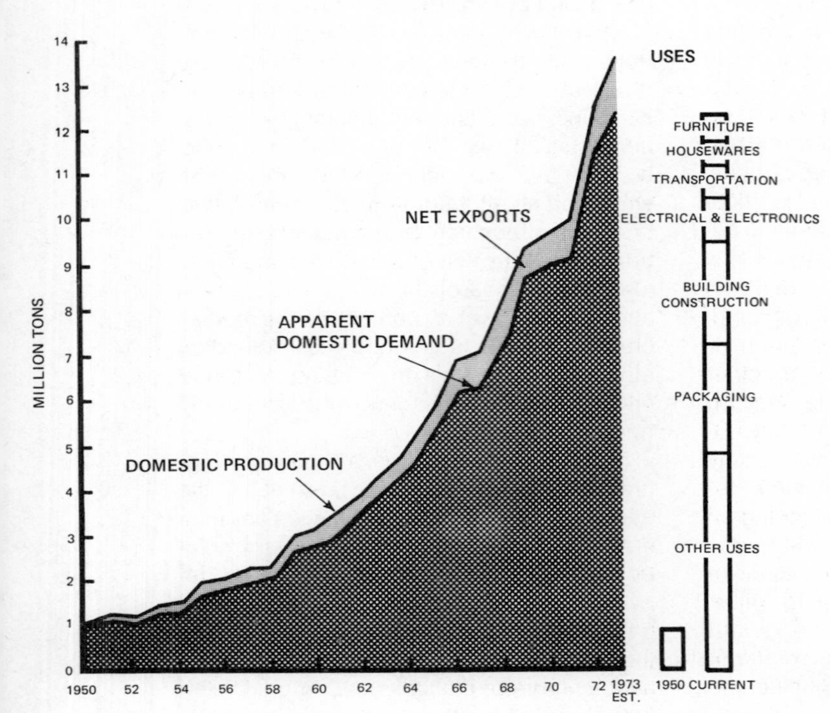

Figure 3. United States supplies and uses of plastics, 1950–73.

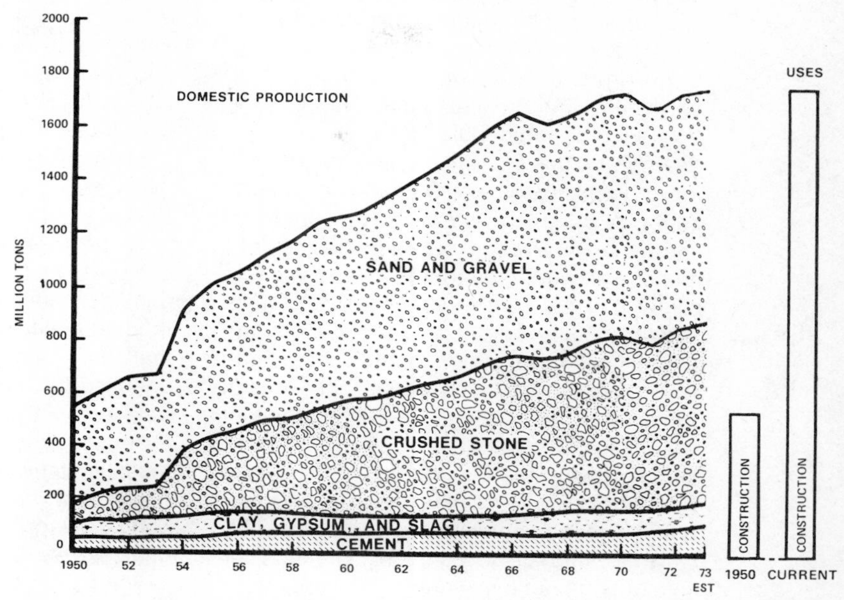

Figure 4. United States supplies and uses of major nonmetallic construction materials, 1950–73.

al demand for nonmetallic minerals. Demand for nonmetallics currently accounts for half of all tonnage of United States mineral demand. The wide use of silicate materials reflects the fact that silicon accounts for about one-fourth and oxygen about one-half of the earth's crust. Concrete, consisting of cement, sand and gravel, is a major construction material which directly competes with metals, as well as requiring use of metal reinforcing in connection therewith. Figure 4 shows the threefold rise in United States use of major nonmetallic construction materials over the past two decades. . . .

As shown by Figure 5, United States demand for energy has doubled over the past two decades. This demand for energy has resulted in relatively steady demand for coal

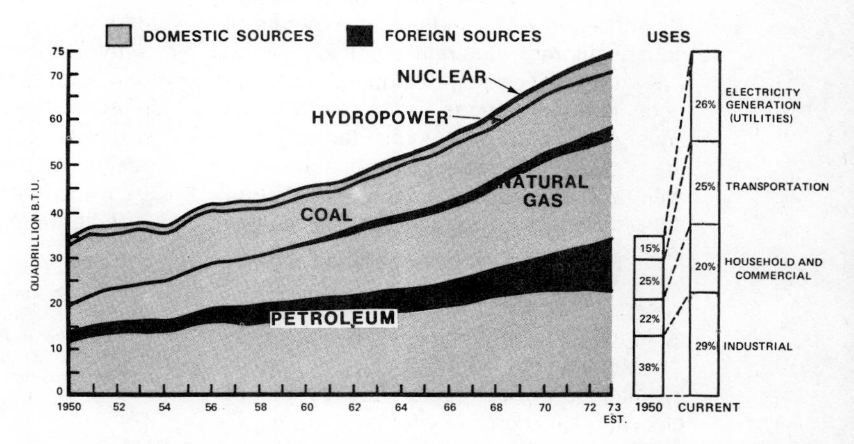

Figure 5. United States demand for major energy sources, 1950–73.

and in significantly increased demand for natural gas and petroleum. . . . In energy, too, the possibilities of substitution are significant, as power plants can be designed to burn oil, coal, or natural gas interchangeably; liquid fuels can be made from agricultural products as well as coal, gas, and petroleum; and gaseous fuels can be made from solid or liquid fuels. Carbon and hydrogen can be linked in many ways to form solids, liquids, or gases. Demand for energy could be substantially modified by presently known conservation possibilities. Many homes and public buildings are too warm in winter and too cold in summer for optimum health. Greater use of insulation, including double windows, storm doors, and shade screening devices will reduce heat loss in winter and heat gain in summer. Smaller vehicles and more use of mass transportation could reduce demand for liquid fuels. . . .

In the grand scheme of nature it would be more appropriate to consider "demand for elements" rather than "demand for minerals" because the living creatures of the earth are inextricably intertwined in the dynamic ecosystem in which the ninety naturally-occurring elements are constantly combining, separating, and recombining in response to fundamental forces. Hematite, magnetite, limonite, siderite, etc., are important minerals, but iron is the more important common elemental constituent of interest. Anthracite, bituminous coal, lignite, peat, and so on are important minerals, but carbon is the most important common elemental constituent of interest. Petroleum and natural gas are important mineral fuels, but carbon and hydrogen are the more important common elemental constituents of interest. Nevertheless, it is still very necessary to consider demand for and supplies of individual minerals because different minerals occur in different ways, different geological, geophysical, and geochemical methods are required to discover them, different beneficiation methods are required to separate and concentrate them, and different chemical and metallurgical processes are required to convert them into more useful forms. Further, many commercial transactions still deal largely with minerals *per se*. For example, columbite, tantalite, ilmenite, rutile, zircon, potash, barite, bauxite, corundum, feldspar, fluorspar, kyanite, magnesite, quartz, talc, vermiculite, etc., are still common items of trade. . . . Production and trade in salt *per se* must be considered because salt is found, mined, processed, and sold as salt. Salt is a major tonnage item, and United States annual demand therefor is almost fifty million tons. Yet it is also important to analyze demand for and supplies of sodium and chlorine, the two elements in common salt, [which are] of particular interest to the chemical industry. . . .

Attempts to assess future needs are normally based on extrapolation of past usage patterns appropriately modified by foreseeable technological, economic, and social changes. In forecasting it is necessary to consider both overall national totals and identifiable components of total demand, so that the validity of projections can be checked by testing extrapolations of gross data against summations of extrapolations for major programs. Projections of demand must lie within the limits of reasonableness. If in recent years, for example, United States annual usage of a certain commodity was one hundred units, it would be unreasonable to forecast near-term future usage as two hundred units or fifty units without clear justification. Two hundred units could be justified if demonstrated major new uses are highly probable. Fifty units could be justified if significant available substitutes are about to be utilized. In the case of materials for which usage is closely tied to defense and aerospace programs, it would not be impossible for near-term future demand to jump from one hundred units to one thousand units or drop to ten units based upon firm programs.

Over the years the Bureau of Mines has made intensive efforts to assess past, present, and future demand. Such studies by the Bureau of Mines have been referred to in the *First* and *Second Annual Reports of the Secretary of the Interior under the Mining and Minerals Policy Act of 1970** . . . When long-term supply-demand trends are projected, future deficits appear evident, giving rise to greater concern as to our future mineral position. Item-by-item quantitative details are provided by Table 2.† . . .

While there are major uncertainties inher-

*U. S. Department of the Interior, *First* and *Second Annual Reports of the Secretary of the Interior under the Mining and Minerals Policy Act of 1970 (P.L. 91–631)* (Washington, D.C., U. S. Government Printing Office). These reports are available at depository libraries.

†The table reprinted here is abridged from the original (Ed.).

TABLE 2. COMPARISON OF PREVAILING U.S. PRIMARY MINERAL SUPPLY-DEMAND WITH PROJECTED HISTORICAL TRENDS IN DOMESTIC MINERAL PRODUCTION (JANUARY 11, 1974)

		1972		2000	
Commodity	**Units‡**	*1972 U.S. Primary Production From Domestic Sources*	*1972 U.S. Primary Demand*	*2000 Projected U.S. Primary Demand*	*2000 U.S. Primary Production If Past 20-Year Trends Prevail*
Aluminum Ores	Thousand S.T.	467	5,083	26,400	551
Arsenic	S.T.	*	*	38,000	1,640
Boron	Thousand S.T.	189	92	303	366
Bromine	Million lb	387	358	500	663
Calcium	Thousand S.T.	87	87	241	126
Chlorine	Thousand S.T.	9,869	9,900	42,300	20,900
Chromium	Thousand S.T.	0	506	1,090	0
Cobalt	Thousand 1b	0	19,268	24,700	0
Copper	Thousand S.T.	1,665	1,951	5,400	2,600
Gold	Thousand T.oz	1,450	7,254	14,300	1,033
Iron	Million S.T.	52	83	153	58
Lead	Thousand S.T.	619	970	1,430	822
Magnesium-Metal	Thousand S.T.	121	108	580	192
Magnesium-Nonmetallic	Thousand S.T.	948	998	2,160	1,370
Manganese	Thousand S.T.	29	1,366	2,360	0
Mercury	Thousand Fl	7	41	57	15
Nickel	Thousand S.T.	17	172	385	38
Nitrogen-Compounds	Thousand S.T.	11,901	11,704	39,700	28,100
Nitrogen-Gas & Liquids	Thousand S.T.	7,011	7,011	20,900	15,900
Palladium	Thousand T.oz	11	713	1,060	30
Phosphorus	Thousand S.T.	5,603	3,963	12,000	12,200
Platinum	Thousand T.oz	5	467	820	0
Potassium	Thousand S.T.	2,207	3,996	12,000	3,750
Silicon	Thousand S.T.	535	561	1,000	950
Silver	Thousand T.oz	37,233	122,257	210,000	45,200
Sodium	Thousand S.T.	18,935	19,834	67,200	37,600
Strontium	S.T.	0	*	34,200	0
Sulfur	Thousand L.T.	10,196	9,833	30,000	15,400
Thorium	S.T.	*	*	1,500	80
Tin	L.T.	*	48,853	90,000	58
Titanium-Metal	Thousand S.T.	0	19	168	0
Titanium-Nonmetallic	Thousand S.T.	228	592	1,810	465
Vanadium	S.T.	5,248	7,209	31,000	10,800
Zinc	Thousand S.T.	478	1,489	3,100	600
Asbestos	Thousand S.T.	132	809	2,400	302
Clays	Million S.T.	60	58	174	82
Garnet	Thousand S.T.	19	16	53	39
Graphite	Thousand S.T.	*	*	108	0
Gypsum	Thousand S.T.	12,328	19,412	35,000	12,200
Mica-Sheet	Thousand lb	14	5,457	600	0
Sand & Gravel	Million S.T.	913	912	3,200	1,700
Stone-Crushed	Million S.T.	707	707	2,520	1,440
Stone-Dimension	Thousand S.T.	1,490	1,615	3,820	656
Talc	Thousand S.T.	1,107	942	2,920	1,710
Anthracite	Thousand S.T.	7,100	5,915	2,300	0
Bituminous Coal & Lignite	Million S.T.	595	520	1,000	818
Natural Gas	Billion C.F.	21,624	22,565	49,000	44,500
Peat	Thousand S.T.	607	917	2.200	1,316
Petroleum†	Million bbl	4,094	5,990	13,500	6,491
Shale Oil	Million bbl	**	**	1,500	0
Uranium	S.T.	11,590	10,250	82,000	20,659

*Certain data are withheld because the manner in which the information became available to the Bureau of Mines does not permit general disclosure.

†Includes natural gas liquids.

**Less than one-tenth unit.

‡S.T. = short ton; L.T. = long ton; T.oz = Troy ounce; C.F. = cubic foot.

ent in any attempt to predict future United States demand for minerals, these difficulties are compounded when efforts are made to predict world-wide demand. The United States has only about 6 percent of the world's population and only about 6 percent of the world's land area. However, the United States now has a demand for about one-quarter of the current annual mineral production of the earth. Obviously, then, the other 94 percent of the people of the earth living on the other 94 percent of the earth's land area are currently using a smaller proportionate share of the earth's annual mineral production. Furthermore, the material standard of living in the United States is higher than in any other country of the world. Indeed, the best measure of a material standard of living is *not* annual new mineral supply but rather the per capita share of use of the *total materials in place*. For example, United States annual new copper "consumption" is reported to be about twenty pounds per person, but the term "consumption" also is a misnomer except from the point of view of the wire mill or brass mill. Most of this new copper is really not "consumed" at all but instead it is converted into copper wire, copper tubing, brass, and alloys from which useful articles having a long life are fabricated. Most of the copper tubing, the copper wiring, the brass hardware, etc., in use now was reported as "consumed" many years ago. But it is estimated that the pool of copper in use in the United States is probably of the order of four hundred pounds per person. Another clear example of the pool of mineral materials in use can be seen in motor vehicles. The more than 120 million motor vehicles registered in the United States constitute a pool of steel in use equivalent to at least two years' full United States demand for steel. When we add up the steel in buildings, railroads, machinery, and so on, it is obvious that the equivalent of many years of annual production is in current everyday use. Some of the material in use could last longer with better maintenance, and some, once discarded, could be more effectively recycled. The same situation applies to almost all mineral materials except the fuels, which are commonly considered to be consumed. However, even fuels are really not "consumed" in many applications but rather the chemical energy originally contained therein is converted to more useful and enduring forms. For example, coal is used to make coke, coke is used to make steel, and the steel then remains in use for long periods of time. Fuel is burned to generate electricity. Substantial quantities of electricity are used to produce metallic aluminum. Much metallic aluminum is then used in enduring applications. Consequently, as we attempt to forecast future demand resulting from the desires of other peoples to attain higher standards of living, we must consider the quantities of minerals that would be involved, not just the annual per capita new increments.

No set of future projections can be guaranteed to be correct. In the past, most projections of mineral demand have tended, if anything, to be too low, and programs based thereon have been too modest. But projections are valuable in indicating the order of magnitude of the future demand-supply situation, because any set of reasonable projections of future United States and world mineral demand make it evident that long-term demands will require substantial additional mineral supplies. Long lead times are required to find mineral resources, to convert them to blocked-out reserves, to mine and concentrate them, to refine them, and to convert them into energy and processed materials of mineral origin, while also conserving, recycling, and rehabilitating. Accordingly, highest priority must be given to closer industry-government-academia cooperation and public cooperation at every stage of the complicated process by which resources are converted into materials and energy to meet the needs of mankind.

8

RUNNING OUT OF FOOD?

THE STAFF OF *NEWSWEEK* MAGAZINE

Perhaps in ten years, millions of people in the poor countries are going to starve to death before our very eyes . . . We shall see them doing so upon our television sets. How soon? How many deaths? Can they be prevented? Can they be minimized? Those are the most important questions in our world today.

When that apocalyptic warning was sounded by British author C. P. Snow [in 1969], it was dismissed by many food experts as unduly alarmist. At that time, miracle seeds and fertilizers were creating a global "green revolution," and there was even talk that such chronically hungry nations as India would soon become self-sufficient in food. But today that sort of optimism is no longer fashionable. World stores of grain are at their lowest level in years—only enough to last for 27 days—and there are grim signs that the current shortage is not just a temporary phenomenon but is likely to get worse.

In the coming decades, some scholars believe, food scarcity will be the normal condition of life on earth—and not only in the poor countries but in the richer ones as well. Unless present trends are somehow reversed, says biologist J. George Harrar, "millions of people in the poor areas will die of starvation. But the affluent societies [including the United States] will experience dramatically reduced standards of living at home." Even Agriculture Secretary Earl Butz, a notorious optimist on the subject of food, concedes that Americans may have to substitute vegetable for animal protein. "We have the technology," Butz told *Newsweek's* Tom Joyce reassuringly, "to make better hamburgers out of soy beans than out of cows."

Even now, food shortages affect the entire world. In the last two years, famine has threatened India and visited widespread misery upon the sub-Sahara nations of Africa where an estimated quarter million people have died. Scarcely less shocking, half of the world's 3.7 billion people live in perpetual hunger. The industrial nations are swiftly buying up the dwindling supplies of food and driving up food prices so high that poorer countries cannot afford to pay them.

Prospects for the future are clouded by the old Malthusian specter of population growth. [By 1975] there will be 4 billion human beings on earth, and by the end of the century that figure is expected nearly to double to 7.2 billion. Food production is simply not growing fast enough to feed that many mouths, and it is unlikely to do so in the decades ahead. A complicating factor in the race between food and people is the burgeoning affluence in such parts of the world as Western Europe, Japan and the Soviet Union. Rising expectations in these areas have bred strong new demands on the world's food supplies. More and more people want their protein in the form of meat rather than vegetables, and this in turn has driven up the need for feed grains for the growing herds of livestock. "Affluence," argues economist Lester Brown, "is emerging as a major new claimant on world food resources."

To meet this proliferating demand for food, insists John Knowles, president of the Rockefeller Foundation, "the world's basic food crops must double in the next eighteen years." The more positive thinkers among the food experts are convinced that this can be done—basically by expanding the area of land under production and by raising the output of crops on the cultivated areas. The world has the means to do the job, they argue—*if* the underproductive countries would order their societies a little better, *if* the richer countries would pump larger amounts of capital and know-how into the less fortunate nations for the development of agriculture, *if* more irrigation and fertilizer were brought into play, *if* mankind would use its common sense.

Many students of the food crisis are far less optimistic. "We have just about run out of good land, and there are tremendous limitations on what we can do in the way of irrigation," contends Prof. Georg Borgstrom of Michigan State University. Economist Brown supports this view. "The people who talk about adding more land are not considering the price," he says. "If you are willing to pay the price, you can farm Mount Everest. But the price would be enormous."

Moreover, Brown and other experts do not expect the sea to solve the world's food problems. Huge fishing fleets have depleted many traditional fishing grounds, and the overall catch is declining. Anchovies, one of the major ingredients in animal feed, recently disappeared from the waters off Peru for two years—largely a result of over-fishing. Water pollution, too, is taking a heavy toll of fish life along the world's continental shelves. And much of the fish that is caught each year is being squandered. "Every year, Americans use tons of tuna fish in pet foods," one food expert points out. "But how much longer will we be able to afford the luxury of feeding our cats and dogs on food people could consume?"

Fertilizer, an essential element, is also becoming prohibitively expensive. Petroleum is a major source of fertilizer, and the towering price of oil thus has a direct effect on agriculture. Dr. Norman Borlaug, sometimes called the "father of the green revolution," has complained bitterly that Arab oil politics, aimed at the industrial countries, will eventually strike most heavily at the developing nations. "India," remarks Brown, "is really up the creek. As a result of the fertilizer shortage, grain production is likely to be off 6 to 9 million metric tons."

On top of all these problems, the world's farmers have been beset by weather conditions that threaten to dislocate food patterns around the world. According to some meteorologists, these changes in climate will probably be a long-range factor. For a variety of reasons, they point out, the earth seems to be cooling off, and this cooling process is causing a southward migration of the monsoon rains. This in turn is producing a dry-weather pattern stretching from the sub-Sahara drought belt through the Middle East to India, South Asia and North China. Even the U.S. could soon be at the mercy of the weather. Some meteorologists are predicting a cyclical return to drought in the Great Plains States—possibly even dust-bowl conditions. "Even a mild drought in this tight supply situation," said one Agriculture Department official, "could be a disaster."

Over the years, the U.S. supplied a staggering $20 billion worth of food to needy countries under Public Law 480, the so-called Food for Peace program. But in recent years, the program has been allowed to wither, and with food demand rising around the world, American farmers—encouraged by the Administration—have flung themselves into the business of exporting food on a strictly cash-and-carry basis. In the fiscal year ending in June 1972, the U.S. exported $8 billion worth of farm products; [in 1973] the figure reached $12.9 billion; and when [fiscal 1974] ends in June it is estimated that it will have zoomed to $20 billion. The U.S. now views agricultural products not as a give away item but as a way of earning the foreign exchange needed to pay for imports, including high-priced crude oil. "Food for crude" is the shorthand for the current policy at the Department of Agriculture.

With virtually all U.S. food surpluses committed to trade, not aid, it is difficult to see how the U.S. can continue to play its old role as provider of food to the world's hungry masses. And there are many people in Washington who do not see this as such a bad thing. "The worst thing we can do for a country," says a State Department official, "is to put it on the permanent dole. That would be an excuse not to solve its own problems, especially population. Now, our thinking is that feeding the world is an international problem, maybe one for the United Nations." That view was underlined [in September of 1973] when Henry Kissinger asked the United Nations to call a world conference on the problems of feeding the world. "No one country can cope with this problem," said the Secretary of State.

In response, the U.N. plans to hold a World Food Conference in Rome [in November of 1974]. Among the major proposals certain to be made are that the less developed nations discourage population growth and that the industrial nations work together to help feed the world's poor. Indeed, Dr. A. H. Boerma, the Dutchman who heads the U.N. Food and Agriculture Organization, has proposed a "world food reserve"—roughly like that of the Biblical Joseph, who advised the Pharaohs to store up grain in good years against future famines. But so far, the suggestion has been greeted with a total lack of enthusiasm in the U.S., Canada and Australia, the only countries in the world with significant food surpluses.

Resistance to an internationally controlled food reserve is easy enough to understand. Farmers fear that such vast stores of controlled food might, at some point, be unloaded on the world market, sending prices down in a dizzying spiral. And governments do not want to give up a formidable political weapon. In the politics of international food, agriculture may very well turn out to be the United States' ace in the hole. "We are not," declares one high-level Washington official, "going to throw that away too easily."

And so, to a very large extent, the U.S., as the greatest food producer in the world, will still be in a position to determine who gets food in the decades ahead; it will almost certainly be American food and American policy that answer the questions posed by C. P. Snow. "We are going to have some big moral decisions to make," says Sen. Hubert Humphrey. "We will be faced with famine situations in Africa, Asia and other parts of the world where there are victims of rising population and bad weather. But the question, I believe, is going to come down to whether Americans will be willing to cut down on their own consumption to help those poor people."

PROGNOSIS

Irving Wender and Bernard D. Blaustein

WEDNESDAY, 9 A.M., LECTURE HALL 12A (ONE WEEK LATER)

Good morning. Now that you've read up on energy and minerals, and the pollution which comes from our uses of them, let me try to put it all into perspective for you and give you my impression of where we're heading and what I think we can do about it.

For some years I collected articles on energy, thinking that a thorough understanding of the onrushing energy problems would help me find solutions for the troubles on the near horizon. This meant also acquiring articles on the environment, on mineral resources which are in short supply, on population trends and on food production—all accompanied by often overriding economic and political considerations. But eventually I was forced to stop stuffing my files with the flood of publications because each journal, magazine and newspaper stated and restated the problem, attempted to lay blame, offered a solution (usually terribly simplistic), predicted shortages and doom, and described promising new technology in terms so glowing that even its inventors scarcely recognized their proposed processes. To put it bluntly, the energy problem is just too complicated to understand and impossible really to explain. It has been said that "much of today's frustration is caused by a surplus of simple answers coupled with a tremendous shortage of simple problems."

You hear so many people ask, "How did it happen? How come this energy crisis is suddenly upon us? Why did we have so little warning? What's wrong with our scientists, economists, politicians, our government—couldn't they see what must have been so obviously roaring down the road?"

Gentle students, we *did* have warnings of the problems we face today. What is happening now has been starting to happen for over 50 years. Let me quote from an address by Charles Walcott to the American Association for the Advancement of Science.* He warned of impending disaster in words which are so familiar today:

> *"The United States' unprecedented growth and her present commanding economic position have been made possible by abundance of natural resources. Individual and public economic policies have been predicated on this abundance. Minerals, forests, fur and game animals, agricultural soils, range lands, fish, and water resources were all seemingly inexhaustible in supply, and all have been appropriated and exploited recklessly and wastefully."*

**Science 184*, 486 (1974).

Walcott said this in 1924, over fifty years ago! And he said it in a time of relative prosperity.

There's one thing we can say with certainty, however: the energy crisis is central to all the others. And there are at least five major aspects of the energy problem: environmental, economic, technological, governmental (or political) and perhaps the most important, what we might call the moral aspect. And they are all interrelated in complex and largely unpredictable ways.

But in spite of the complexity of the problem, each of us has his or her own favorite, personal way of viewing the effect of the energy crisis on our economy and on the world situation in general. There are certain practical questions we all wonder about. The Arab oil embargo came and went; will it come again? Will I be able to get gasoline this summer on my vacation? Heating oil next winter? What will they cost? Will the U.S. be able to afford the huge imbalance of payments for fuels and minerals, which could reach 50 billion dollars annually before the end of the century? Will other countries that control vital commodities (we are short of bauxite for aluminum, chromium, tin, zinc, copper and a host of other minerals) also form cartels to control prices and supply the way the oil producers have done? Does this country have the capital, the materials, the water and other natural resources, the trained manpower and so on to carry out the proposed research and development programs in coal gasification and coal liquefaction, in shale oil and tar sand development, in research on geothermal, solar and fusion energy sources? Can we harness these new sources of energy without swamping the environment? Do we have the moral fiber to stand up to what Stuart Udall has termed "a revolution in values?"

More specifically, will atomic energy, that nuclear knight on the white horse, save our skins? It doesn't seem so to me, although it will certainly be of tremendous help. After some 30 years and enormous amounts of money and effort, nuclear energy supplied only some 1 percent of our total consumption in 1973. And our largest single energy research and development program, the breeder reactor, may well pose more problems than its eventual worth. It will manufacture large amounts of plutonium, an extremely toxic element that, if spilled or inadvertently dispersed, would be around to haunt (and kill) us for thousands of years;* and it is not too hard to imagine terrorists hijacking enough plutonium to fashion an atomic bomb that would dwarf any previous such threat to international sanity. And yet one must ask, can we manage *without* the breeder reactor?

Well, I didn't promise you any solutions. And in a couple of lectures I can't even point out all the energy problems, their interrelationships and possible solutions. But there is one particular course of action I *must* emphasize and recommend to you. It is so important, of such widespread and long-lived usefulness, so basic to our existence, so morally right, so central to every thought involved, and really not terribly difficult to carry

*See "Environmental Hazards of Nuclear Wastes", page 145 (Ed.)

out, that I'll skip all other avenues of thought and recommended actions: CONSERVATION of our energy, mineral and other natural resources.

When the concept of energy conservation burst upon us following the 1973 Arab-Israeli war, we became acutely aware of how profligate we had been. We lowered our thermostats in the winter and raised them in the summer (those of us, at least, who could figure that out), we turned out unnecessary lights, we drove at lower speeds (some friend or relative is now alive because of this), we conserved gasoline by driving less and by forming car pools, and we started to ride our bicycles seriously. Americans began to think about energy savings in terms of what they could do as individuals, and of how it affected their individual ways of life. And industry, the major consumer of energy, also started to respond and made some very significant savings.

But it didn't last long. When the oil embargo was lifted we seemed suddenly to revert to our old ways. The roads again were full of cars burning more expensive gasoline, and the 55 mile per hour speed limit was pretty much ignored. Sure, we're still worried, but the energy crisis was just one of those things that came and went. But a new time of reckoning is bound to come, in which we'll have to pay more, often much more, for what we use. The solution—the *only* solution—has to be for each of us to try to change his or her way of life a bit so as to conserve energy and minerals as much as possible. The fundamental choice is either a serious commitment to conservation or a breakneck dash to develop new energy sources regardless of their effect on the environment at home or on our political interests abroad.

Yes, we must change our values, though their roots lie deep. If we wish to survive in dignity and not as ciphers in some vast ant-heap society, we must reassume our full moral responsibility. The earth is not just a banquet at which the people of the United States are free to gorge on a third of the world's energy. There is no fundamental scarcity in Nature. It is only our numbers and, above all, our wants that have outdistanced Nature's bounty. We become rich precisely in proportion to the degree in which we eliminate greed and pride from our lives; but as history shows this is not a principle easily learned by humanity. Nevertheless, we must choose this course, for there is no other way to defeat the gathering forces of scarcity.

How? In one of the first reports of its kind, the Office of Emergency Preparedness showed recently that the United States could reduce its energy demands by more than 7 million barrels of oil a day by 1980 if it practiced conservation. That much oil would be worth about $11 billion per year at mid-1973 prices; it would be only speculation to estimate the savings by 1980, but they would be huge. The report declared that "the most significant energy conservation measures are the installation of improved insulation in both new and old homes, the use of more efficient air conditioners, a shift of intercity freight from trucks to rail, intercity passengers from air to rail and bus, and urban passengers from automobiles to motorized mass transit, and the introduction of more efficient equipment and processes." Some lighting experts say there is far too much light in many modern buildings. Office air conditioning often is needed to remove the heat caused by indoor lighting, even when it's snowing outside! Display lighting often burns throughout the night, when most people are sleeping.

To alleviate our energy needs, I'd like to make the following recommendations:

(1) Education: to teach the consequences of conservation and its impact on present lifestyles.

(2) The elimination of enticements to use more, such as declining prices for higher energy usage.

(3) Recycling: minimize "throw-away" or disposable items.

(4) Switch to systems that use energy less intensively.

(5) Emphasize efficiency in the utilization of energy.

(6) Use better insulation and heat-exchange systems.

(7) Use mass transit and car pools.

(8) Run short-and long term cost/benefit analyses on proposed new courses of action and/or regulations. Include social, environmental, and economic considerations.

(9) Allocate and/or ration selectively, if marketing systems and reasonable incentives fail.

(10) Implement a national energy policy that would direct various fuels to "superior" end uses. These would include: (a) petroleum for transportation and as a chemical raw material, (b) natural gas for residential and commercial space heating, (c) coal and nuclear energy for industrial use and the generation of electricity.

The concept of recycling materials is familiar to all of you. We are learning how to reuse glass bottles, to recycle aluminum from beer and soda pop cans, to burn organic wastes to obtain energy, and to make oil or gas from these same wastes. So why do we have to conserve? Why can't we just keep using our energy and minerals over and over again?

The answer will surprise, and perhaps even shock you: energy, unlike other resources, cannot be recycled! It's not just difficult. It's impossible. Once used, energy is essentially gone forever.

How come? Let me explain.

The First Law of Thermodynamics (there are three) is also called the Law of Conservation of Energy. It says simply that energy cannot actually be consumed (destroyed) at all. So when we say that energy is "consumed", we can mean only that it is being changed in form.

"O.K.", you say, "We can't destroy it. So much the easier for us to recycle it, right?" Wrong. It's the Second Law of Thermodynamics that really does us in. Because *it* says that changes of energy from one form to another proceed in such a way as to reduce, on the whole, the availability of energy to do work; and if you can't do work with it, what good is it? To put it differently, it's the *usefulness* of energy that is effectively consumed, even though the energy itself is still present in one form or another. Energy that has been used to do work is said to have been "degraded". For example, most of the electrical energy used to light a bulb is degraded to heat. In turn, the light energy which is emitted may be used to do some work, such as making a photograph or operating a photoelectric cell, but in the process it'll be degraded further.

We can go the other way, but the Second Law still holds. That is, we can turn energy from a less useful into a more useful form, but only at the expense of degrading part of the energy we started with. Such a process is called "energy conversion", and the generation of electricity is an example. If we burn coal or oil to produce electricity, the fire's heat energy is partly converted into electrical energy (which is more useful to us because we can transmit it over long distances) and partly degraded into practically useless low-temperature heat. Thus, the Second Law—a law not of man but of Nature—tells us that no matter what we do or what we'd like to do—converting water power or wind or coal or uranium atoms into heat or electricity until we're blue in the face—the capacity to do work which is contained in any given amount of energy can be used only once. As far as energy is concerned, it's downhill all the way. You can't break even. You can't recycle. *Now* do you believe that conservation is the only way to go?

So by all means recycle your beer cans and bottles and newspapers and all the million-and-one other things and gadgets and junk we affluent Americans have amassed around us. But *conserve* your share of precious energy. Your days of splashing around recklessly in a pool of limitless energy are finished. Prepare to change your lifestyle!

From the film RECYCLING by Stuart Finley, Inc. Supplied through the courtesy of Science.

SECTION TWO

A GLUT OF STUFF AND THINGS

Interpress Film "Atte

The Problem

In 1951, the American Chemical Society celebrated the 75th anniversary of its founding. President Harry S Truman sent his official greetings, a commemorative stamp (of three-cent denomination in those dear, dead days!) was issued, there was the predictable round of symposia and dinners, and a Diamond Jubilee Volume was published. With the centennial of the Society and the bicentennial of the nation close at hand, that large, blue-bound book makes interesting reading. Understandably self-congratulatory, it is chiefly devoted to summarizing progress during the first 75 years of officially organized American chemistry.

And, indeed, there was much progress to report in 1951. The membership of the Society had grown from 230 in 1876 to 63,349; its expenditures had risen from $417 to $2,840,131; and the number of editorial pages printed had increased from 132 to 21,524. Outside the ACS, changes of comparable magnitude had occurred. The population of the United States had swelled from 39.8 million to 150.7 million as the average life span rose from 40 years to 67.2 years and the death rate dropped by one-half. The national income had mushroomed; so had federal expenditures and the national debt. Contributing to this expanding affluence were large increases in the production of wheat, corn, and cotton; the consumption of fertilizer; and the utilization of coal, petroleum, and natural gas. And chemistry had played a significant role in all of this.

Thus, it is easy to see why the publishers of the Diamond Jubilee Volume chose to entitle it *Chemistry . . . Key to Better Living.* The authors of the various sections could point with justifiable pride to great advances in the understanding of chemical phenomena—contributions which, by 1951, had elevated the scientific stature of the American Chemical Society and American chemists to the highest rank. But, beyond that, the writers of 25 years ago could take equal satisfaction in the practical application of that new knowledge. After all, the increase in life span was in no small measure a consequence of penicillin, sulfanilamide, and hundreds of other wonder drugs. A new and powerful insecticide, DDT, had recently saved millions all over the world from the scourge of malaria and typhus. Moreover, DDT, herbicides such as 2,4-D, and chemical fertilizers were in large measure responsible for raising crop yields to unprecedented levels. And the convenience, perhaps even the quality, of life was enhanced daily by creations of the laboratory: nylon, orlon, and similar "miracle fibers"; plastics and synthetic rubber, which outperformed and outlasted natural products; new detergents superior to soap; and a host of other products undreamed of by the handful of visionaries who met on April 6, 1876 to organize the American Chemical Society. No wonder that N. Howell Furman, President of the Society in 1951, could write: "There is every reason to believe that our present slogan, 'Chemistry . . . Key to Better Living' will be justified in myriads of ways in the next quarter century."

That quarter century is now up. And I, for one, believe that Furman was right, though perhaps not precisely in the way he originally intended his optimistic prediction. The benefits of chemistry are still very much with us. In many respects, progress in basic and applied chemistry since 1951 has surpassed the remarkable growth of the previous 75 years. But the contributions of chemistry are by now so much a part of our lives that we seldom reflect just how dependent we are upon them. If anything, we tend to underrate our debt to chemical technology because we have recently discovered with chagrin the magnitude of the payment which chemical technology has already exacted from us. We have paid with unbreatheable air and undrinkable water. And we have paid with dead songbirds and diseased children.

In no small measure, our growing awareness of the extent of the environmental impact of misapplied chemistry is a response to the eloquent voice of Barry Commoner. Therefore, it is appropriate that the first selection in this Section (Article 9) is excerpted from his influential book, *The Closing Circle*. Dr. Commoner is, first and foremost, a scientist with excellent academic credentials—B.A., Columbia; M.A. and Ph.D., Harvard. As a Professor at Washington University, his chief research efforts have been directed at investigating the physiochemical bases of biological processes. Moreover, his active role in the Scientists' Institute for Public Information and the St. Louis Committee for Environmental Information testify that he has thoroughly explored the boundary where science and society meet.

In the first part of the selection reprinted here, Dr. Commoner reviews some of the great post-World War II achievements of science and technology, particularly of chemistry. But he ends on a disquieting note as he asks, "Is it possible that the new technology is a major cause of the environmental crisis?" That the answer is "yes" is the central thesis of *The Closing Circle.* Dr. Commoner argues that the spectacular success of technology in solving short-range problems while creating more serious long-range hazards is a more significant contributor to environmental deterioration than an exploding global population or the expanding affluence of the industrialized nations.

In his book, Professor Commoner goes on to provide several case studies which support his central theme. Specifically, he considers fertilizers, pesticides, detergents, and in the second part of the selection included here, synthetic fibers. To be sure, the resistance of these man-made polymeric materials to biological degradation gives them great practical advantages over natural fibers which mildew and decay. But simultaneously, these same characteristics create a disposal problem of colossal magnitude.

The unhappy prospect of this beautiful blue-green ball as a galactic garbage dump is the message of Article 10, by Wesley Marx. It is a chapter in a book appropriately entitled *Man and His Environment: Waste.* Mr. Marx, an author and journalist specializing in environmental subjects, indicts modern man and his technology as the chief perpetrators of our current global mess. "Aboriginal hunters must rank as cleaner creatures than we, simply because they were unable to dirty their living spaces to the degree we can. While we deodorize our armpits, our wasteloads overcome the waste-receiving capacity of the natural environment in a rather indiscriminate form of chemical and biological warfare."

We have used many weapons against Nature in this war. Ironically, most of them were first introduced to enhance our existence. Mankind has been long aware of potential dangers associated with some of the substances employed. For example, the toxicity of lead was well known to the ancients. It has even been suggested that the decline and fall of the Roman Empire was hastened by lead poisoning. Be that as it may, the industrial use of lead has grown with our technology. To be sure, the employment of lead pigments in paints has been discontinued, but lead-containing paint still peels from tenement walls and finds its way into the bodies of innocent and unsuspecting ghetto children. A far more widely dispersed form of lead is a product of the age of the automobile. Lead tetraethyl, $Pb(C_2H_5)_4$, was introduced in 1923 with the best of intentions—to increase the octane rating of gasoline and thus prevent engine "knocking" due to preignition. Today, lead from the exhausts of millions of cars is literally spread from pole to pole. And there is a danger that short supplies of petroleum may tempt manufacturers and consumers to tolerate higher levels of lead in gasoline.

In Article 11, reprinted from *Environment*, Paul P. Craig and Edward Berlin summarize some of the biochemical effects of this toxic burden. The collaboration of these two authors is noteworthy because it is indicative of the interdisciplinary complexity of such ecologically-related problems. Dr. Craig is a physicist at the Brookhaven National Laboratory and Chairman of the Committee on Environmental Impact of the Automobile, an organization of the Environmental Defense Fund. Mr. Berlin is an attorney, a senior partner in the Washington law firm of Berlin, Roisman, and Kessler. "Existing knowledge," the authors conclude, "demands decisive, expeditious action." Clearly, it also demands the attention and expertise of both the scientific and legal communities and the consent of an informed society.

As Neville Grant implies in Article 12, mercury is at the center of a similarly complex crisis. Dr. Grant is Assistant Professor of Clinical Medicine at the Washington University School of Medicine. As an internist, he is well qualified to comment on the physiological fate of mercury. But his paper also describes some of the tortuous paths through which the element can enter the human ecosystem. Most of the sources are again directly linked to applied chemistry. There may be a temptation to dismiss the story of the Ernest Huckleby family as an unfortunate accident: the chance contamination of pork and people with a methyl mercury fungicide originally applied to millet seed. It almost seems as specific and as potentially controllable as the poisoning of hatters, maddened as they removed hair from hides treated with mercury compounds. But paper mills and chlor-alkali plants spread their effluents far more indiscriminately. Although the latter factories account for the largest commerical use of mercury, they should ideally produce no waste metal. The silvery liquid serves as a cathode for the electrolysis of molten sodium chloride to yield sodium and chlorine, two high-volume industrial chemicals. The reduced sodium metal dissolves in the mercury to form an amalgam. When water is added, the sodium reacts to yield sodium hydroxide, and the mercury is reclaimed—theoretically, in 100 percent yield. But as any student of chemistry knows, 100 percent yields are rare. The presence of mercury in Minamata Bay tuna and Lake Erie pickerel attests that the recovery is not complete. The toxic element has begun its slow path up the ecological food chain. A discussion of the biochemical complexity of that chain forms part of Dr. Grant's article.

Among the many chemical substances which are useful and widely employed in manufacturing technology are the polychlorinated biphenyls (PCBs), a class of compounds chemically similar to the insecticide DDT (see Section Three, page 113) and which, like DDT, has recently been found to constitute a hazard of unexpected magnitude. PCBs are used as plasticizers to make polymers more flexible; as heat transfer media, hydraulic fluids, lubricants, and dielectrics in condensers; and as constituents of microencapsulated "carbonless carbon paper." Hence, PCBs are widely dispersed throughout our technological society. The findings reported in Article 13 by Kevin P. Shea, a fellow editor of this book and an editor of *Environment*, indicate that these compounds are also a fairly common contaminant of the food we eat. In some instances, for example that of "Yusho disease," the contamination has been painstakingly traced to industrial mistakes. Hopefully, recent restrictions on the use and disposal of PCBs will reduce the chances of such incidents in the future. But as Shea concludes, "because of the enormous amounts of PCBs already circulating in the environment, it will be some time before they will be considered insignificant contaminants."

What actually constitutes a contaminant is, of course, often a difficult matter of definition. Perhaps nowhere is the problem more evident than in the case of food additives. These are chemicals intentionally added to foodstuffs, almost always for good purposes—to increase their resistance to decay, mold, and rancidity; to improve flavor, texture, color, and appearance; and to enhance nutritional value. Yet, to some critics, all food additives are food poisons. And indeed, recent experience with monosodium glutamate and cyclamates suggest that we exercise extreme caution in what we add to what we eat. Tom Alexander, an associate editor of *Fortune,* considers some of the nutritional, technical and economic issues in Article 14, which was one of a series on "embattled consumer products." *Fortune* is hardly a scientific journal; nevertheless, it is quite appropriate that a selection from that magazine should appear in a book that attempts to explore the complexity of the interaction between science and society.

This same theme permeates the final article in this Section. And its author is eminently qualified to comment on it. A distinguished electrical engineer, Jerome B. Wiesner has spent much of his life at the science-society interface. During World War II, he worked on the development of radar in the Radiation Laboratory of the Massachusetts Institute of Technology. He subsequently held a number of important administrative offices at M.I.T., culminating in his being named President of that great university in 1971. And, as Science Advisor to President John F. Kennedy, he brought his counsel to the most politically powerful office in the world. It is evident in his article, reprinted here from *Technology Review,* that Dr. Wiesner is no unquestioning apologist for science and technology. Nevertheless, he does conclude that "Technology is for Mankind," a message which must not be overlooked in evaluating the papers which have preceded his.

I now gladly surrender my forum to the authors of these articles, letting them speak for themselves. They speak with fervor and conviction, well supported with scientific data and possessed by a common concern for the human condition.

9
POPULATION AND AFFLUENCE

BARRY COMMONER

The last fifty years have seen a sweeping revolution in science, which has generated powerful changes in technology and in its application to industry, agriculture, transportation, and communication. World War II is a decisive turning point in this historical transition. The twenty-five years preceding the war is the main period of the sweeping modern revolution in basic science, especially in physics and chemistry, upon which so much of the new productive technology is based. In the approximate period of the war itself, under the pressure of military demands, much of the new scientific knowledge was rapidly converted into new technologies and productive enterprises. Since the war, the technologies have rapidly transformed the nature of industrial and agricultural production. The period of World War II is, therefore, a great divide between the scientific evolution that preceded it and the technological revolution that followed it.

We can find important clues to the development of postwar technology in the nature of the prewar scientific revolution. Beginning in the 1920's, physics broke away from the ideas that had dominated the field since Newton's time. Spurred by discoveries about the properties of atoms, a wholly new conception of the nature of matter was developed. Experiment and theory advanced until physicists gained a remarkably effective understanding of the properties of subatomic particles and of the ways in which they interact to generate the properties of the atom as a whole. This new knowledge produced new, more powerful techniques for smashing the hitherto indestructible atom, driving out of its nucleus extremely energetic particles. Natural and artifical radioactivity was discovered. By the late 1930's it became clear, on theoretical grounds, that vast quantities of energy could be released from the atomic nucleus. During World War II, this theory was converted into practice, giving rise to nuclear weapons and reactors—and to the hazards of artificial radioactivity and the potential for catastrophic war.

The new physical theories also helped to explain the behavior of electrons, especially in solids—knowledge that in the postwar years led to the invention of the transistor and to the proliferating solid-state electronic components. This provided the technological base for the modern computer, not to speak of the transistor radio.

Chemistry, too, made remarkable progress in the prewar period. Particularly significant for later environmental effects were advances in the chemistry of organic compounds. These substances were first discovered by eighteenth-century chemists in the juices of living things. Gradually chemists learned the molecular composition of some of the simpler varieties of natural organic substances. Chemists developed a powerful desire to imitate nature—to synthesize in the laboratory the organic substances uniquely produced by life.

The first man-made organic substance, urea, was synthesized in 1828. From this simple beginning (urea contains only one carbon atom), chemists learned how to make laboratory replicas of increasingly complex natural products. Once techniques for putting together organic molecules were worked out, an enormous variety of different products could be made. This is the natural consequence of the escalating mathematics of the possible atomic combinations in organic compounds. For example, although molecules that are classed as sugars contain only three types of atoms—carbon, oxygen, and hydrogen—which can be related to each other in only a few different ways, there are sixteen different molecular arrangements for sugars that contain six carbons (one of these is the familiar glucose, or grape sugar). The number of different kinds of organic molecules that can, in theory, exist is so large as to have no meaningful limit.

Around the turn of the century chemists learned a number of practical ways of creating many of the theoretically possible molecular arrangements. This knowledge—that the variety of possible organic compounds is essentially limitless and that ways of achieving at least some of the possible combinations were at hand—proved irresistible. It was as though language had suddenly been invented, followed inevitably by a vast outburst of creative writing. Instead of new poems, the chemists created new molecules. Like some poems, some of the new molcules were simply the concrete end-product of the joyful process of creation—testimony to what the chemist had learned. Other molecules, again like certain poems were created for the sake of what they taught the creator—newly defined steps toward more difficult creations. Finally, there were new molecules created with a particular purpose in mind—let us say to color a fabric—the analogy, perhaps, of an advertising jingle.

The net result represents, in terms of the number of new man-made objects, probably the most rapid burst of creativity in human history. Acceleration was built into the process, for each newly created molecule became, itself, the starting point for building many new ones.

As a result, there accumulated on the chemists' shelves a huge array of new substances, similar to the natural materials of life in that they were based on the chemistry of carbon, but most of them absent from the realm of living things. As new useful materials were sought, some of the chemicals were taken off the shelf—either because of a resemblance to some natural substance or at random—and tried out in practice. This is how sulfanilamide—which a dyestuff chemist had synthesized in 1908—was found in 1935 to kill bacteria, and DDT—which had sat on a Swiss chemical laboratory shelf since 1874—was found in 1939 to kill insects.

Meanwhile a good deal was learned about the chemical basis of important molecular properties—the kind of molecular structure that governed a substance's color, elasticity, fibrous strength, or its ability to kill bacteria, insects, or weeds. It then became possible to design new molecules for a particular purpose, rather than search the chemical storeroom for likely candidates. Although many such advances had occurred by the time of World War II, very few had as yet been converted to industrial practice on a significant scale. That came later.

Thus, the prewar scientific revolution produced, in modern physics and chemistry, sciences capable of manipulating nature—of creating, for the first time on earth, wholly new forms of matter. But until World War II the practical consequences were slight compared to the size and richness of the accumulated store of knowledge. What the physicists had learned about atomic structure appeared outside the laboratory only in a few kinds of electrical equipment, such as certain lamps and x-ray apparatus. In industry, physical phenomena still appeared largely in the form of mechanical motion, electricity, heat, and light. In the same way the chemical industry was largely based on the older substances—minerals and other inorganic chemicals. But the new tools, unprecedented in their power and sweeping in their novelty, were there, waiting only on the urgency of the war and the stimulus of postwar reconstruction to be set to work. Only later was the potentially fatal flaw in the scientific foundation of the new technology discovered. It was like a two-legged stool: well founded in physics and chemistry, but flawed by a missing third leg—the biology of the environment.

All this is a useful guide in our search for the causes of the environmental crisis. Is it only a coincidence that in the years following World War II there was not only a great outburst of technological innovation, but also an equally large upsurge in environmental pollution? *Is it possible that the new technology is the major cause of the environmental crisis?*

THE TECHNOLOGICAL FLAW

BARRY COMMONER

Textile production reflects [an] important displacement of natural organic materials by unnatural synthetic ones. Some relevant statistics: in 1950 in the United States about 45 pounds of fiber were used per capita by fabric mills. Of this total, cotton and wool accounted for about 35 pounds per capita, modified cellulose fibers (such as rayon) for about 9 pounds per capita, and wholly man-made synthetic fibers (such as nylon) for about 1 pound per capita. In 1968 total fiber consumption was 49 pounds per capita. However, now, cotton and wool accounted for 22 pounds per capita, modified cellulose fibers for 9 pounds per capita, synthetic fibers for 18 pounds per capita. "Affluence"—per capita use of fiber—was essentially unchanged, but natural materials had been considerably displaced by synthetic ones. This technological displacement has intensified the stress on the environment.

To produce fiber, whether natural or synthetic, both raw materials and a source of energy are required. The molecules that make up a fiber are themselves threadlike polymers—chains of repeated smaller units. In cotton the polymer is cellulose, long molecules composed of hundreds of glucose units linked end to end. Energy is needed to assemble such an elaborate structure—both to form the necessary glucose units and to join them into the molecular thread. The energy required to form the cotton fiber is, of course, taken up by the cotton plant from a free, *renewable* resource—sunlight. The energy needed to form wool, which is made up of the protein polymer keratin, is obtained from the sheep's food, which is, in turn, derived from sunlight.

The crucial link between an energetic process and the environment is the temperature at which the process operates. Living things do their energetic business without heating up the air or polluting it with noxious combustion products. Whether in the cotton plant or the sheep, the chemical reactions that put the natural polymers together operate at rather low temperatures, and the energy is transferred efficiently. Nothing is burned, nothing wasted. This is, after all, the hallmark of life—that it can function well in the earth's environment, at the temperatures common on the planet. The chemical composition of a complex system such as the earth's surface is determined by its general temperature, for as the temperature of a system is raised, otherwise inactive constituents can react, thereby changing its chemical composition. For example, at the range of temperatures on the earth, chemical reaction between oxygen and nitrogen is negligible, so that the existence of these gases in the air, and the absence of their reaction products, such as nitrogen oxides, reflects the earth's temperature range. Thus, nitrogen oxides are rare at the earth's general temperature, even though the ingredients—oxygen and nitrogen gas—are intimately mixed in large amounts in the air—because these two gases do not react appreciably except at much higher temperatures. Living things, having evolved in the absence of nitrogen oxides, have developed a number of chemical processes that would be damaged if nitrogen oxides were present. As a result, living things are susceptible to toxic effects from nitrogen oxides and are therefore at risk whenever abnormally high temperatures occur and nitrogen oxides are produced.

The energy required for the synthesis of a synthetic fiber like nylon comes from two sources. Part of it is contained in the raw materials; since these are usually derived from petroleum or natural gas, their energy represents solar energy previously trapped by fossil plants. This is, of course, a nonrenewable source of energy and therefore, in the ecological sense, wasteful.

Another part of the energy used in nylon synthesis is needed to separate the various raw materials from petroleum or natural gas and to drive the various chemical reactions. Nylon, for example, is produced by a series of from six to ten chemical reactions, operating at temperatures ranging from 200°F (near the boiling point of water) to 700°F (about the melting point of lead). This means a considerable high-temperature combustion of fuel—and, for the reasons given, the resulting air pollution. Such reactions may release waste chemicals into the air or water, producing an environmental impact not incurred in the production of a natural fiber.

Of course the production of cotton or wool can also violate ecological principles, and does as presently practiced. In the United States cotton is now grown with intensive application of nitrogen fertilizer, insecticides, and pesticides, all of which produce serious environmental impacts that are avoided in the manufacture of synthetic fibers. In addition, the gasoline burned by tractors engaged in cotton production leads to air pollution. Some of these effects could be minimized; for example, more reliance could be placed on natural control of insect pests. Similarly nylon production could be improved, ecologically, by reducing waste chemical emissions. However, what is at issue here is the fundamental point that even if all possible ecological improvements were made in the two processes, the natural one would still be more advantageous ecologically. Cotton involves a freely available, nonpolluting, renewable source of energy—sunlight—for the basic chemical synthesis, whereas in the chemical synthesis of a fiber, energy must be derived from a nonrenewable resource, and through high-temperature operations that emit ecologically harmful wastes. Even with the best possible controls such operations pollute the environment with waste heat.

Once produced, a synthetic fiber inevitably generates a greater impact on the environment than a natural fiber. Because it is man-made and unnatural, the synthetic fiber is not disposable without stressing the environment. On the other hand, the natural polymers in cotton and wool—cellulose and keratin—are important constituents of the soil ecosystem and therefore cannot accumulate as wastes if returned to the soil.

The ecological fate of cellulose, whether

in a leaf, a cotton shirt, or a bit of paper, is well known. If it falls on the ground and becomes covered with soil, such cellulosic material enters into a series of complex biological processes. The cellulose structure is first invaded by molds; their cellulose-digesting enzymes release the constituent sugars into the soil. These stimulate the growth of bacteria. At the same time, the degradation of cellulose allows enzymatic attack on other polymeric components of the leaf, releasing soluble nitrogenous constituents to the soil. These too stimulate bacterial growth. The net result is the development of fresh microbial organic matter, which becomes converted to humus—a substance essential to the natural fertility of the soil. Because cellulose is an essential cog in the soil's ecological machinery, it simply cannot accumulate as a "waste." Keratin behaves similarly in the soil ecosystem. All this results from the crucial fact that for every polymer produced in nature by living things, there exist enzymes that have the specific capability of degrading that polymer. In the absence of such enzymes, the natural polymers are quite resistant to degradation, as is evident from the durability of fabrics protected from biological attack.

The contrast with synthetic fibers is striking. The structure of nylon and similar synthetic polymers is a human invention and does not occur in natural living things. Hence, unlike natural polymers, synthetic ones find no counterpart in the armamentarium of degradative enzymes in nature. Ecologically, synthetic polymers are literally indestructible. And, as in the case of natural polymers, there are no other natural agencies capable of degrading polymers at a significant rate. Hence, every bit of synthetic fiber or polymer that has been produced on the earth is either destroyed by burning—and thereby pollutes the air—or accumulates as rubbish.

This fact is apparent to anyone who has wandered along a beach in recent years and marveled at the array of plastic objects cast ashore. A close look at such objects—bits of nylon cordage, discarded beer-can packs, and plastic bottles—is even more revealing. Like other objects on the beach, bits of glass for example, the plastic objects are worn by wave action. Ecologically, it is always useful to ask about any given material in the environment, "Where does it go?" Where, then, does the material abraded from plastic objects go in the marine environment? The answer has just become apparent in a recent report. Nets which are used to collect microscopic organisms from the sea, now accumulate a new material: tiny fragments of plastic fibers, often red, blue, or orange. For technological displacement has been at work here; in recent years natural fibers such as hemp and jute have been nearly totally replaced by synthetic fibers in fishing lines and nets. While the natural fibers are subject to microbial decay, the synthetic ones are not and therefore accumulate.

It is illuminating, in this connection, to ask why synthetic cordage has replaced natural materials in fishing operations. A chief reason is that the synthetic fibers have the advantage of resisting degradation by molds, which, as already indicated, readily attack cellulosic materials such as hemp or jute. Thus, the property that enhances the economic value of the synthetic fiber over the natural one—its resistance to biological degradation—is precisely the property that increases the environmental impact of the synthetic material.

All modern plastics, like synthetic fibers, are composed of man-made, unnatural polymers. They are, therefore, ecologically nondegradable. It is sobering to contemplate the fate of the billions of pounds of plastic already produced. Some of it has of course been burned—thereby adding to the air not only the ordinary products of combustion, but in some cases particular toxic substances such as hydrochloric acid as well. The rest remains, in some form, somewhere on the earth.

Having been designed for their plasticity, the synthetic polymers are easily formed into almost any wanted shape or configuration. Huge numbers of chaotically varied plastic objects have been produced. Apart from the aesthetic consequences, there are serious ecological ones. As the ecosphere is increasingly cluttered with plastic objects nearly infinite in their shape and size, they will—through the workings of nature and the laws of probability—find their way into increasingly narrow nooks and crannies in the natural world. This situation has been poignantly symbolized by a recent photograph of a wild duck, its neck garlanded with a plastic beer-can pack. Consider the awesome improbability of this event. A particular plastic pack is formed in a factory, shipped to a brewer, fitted around six cans of beer, further transported until it reaches human hands that separate plastic from beer can. Then, tossed aside, it nevertheless persists until it comes to float on some woodland lake where a wild duck, too trustingly innocent of modern technology, plunges its head into the plastic noose. Such events—which bring into improbable, wildly incongruous, but fatal conjunction some plastic object and some unwitting creature of the earth—can only become increasingly frequent as plastic factories continue to emit their endless stream of indestructible objects, each predestined by its triumphant escape from the limited life of natural materials, to become waste.

10
MAN AS WASTE-MAKER

WESLEY MARX

Waste-free activities are hard to find. Activity itself dissipates energy and matter, thus eating, working, playing, farming, and warring all have one common trait—debris. If we are going to exist as more than sightseers and lovers, we are going to produce wastes.

Man is not alone in his waste-making. All life depends on dissipated energy and materials for survival. A blade of grass respires gases into the atmosphere during photosynthesis and provides us with oxygen. Under the warm sun, lakes and oceans release water vapor into the atmosphere to form clouds that irrigate the earth and, in turn, replenish the lakes and the oceans. Frost and wind grind down boulders and release grains of quartz into mountain streams. These grains, ultimately transported to the ocean littoral, accrete on the shore to form 5-mile-long beaches and 200-foot-high sand dunes.

If dead plants and animals remained inert, life would have long since crowded itself off the planet. However, death is integrated into the metabolism of the environment as effectively as respired gases and evaporated water. Dead plants and animals decay and decompose as if they had been outfitted with self-destruct systems. But reconversion, not destruction, is taking place. The forces of decomposition, from vultures to bacteria, break down the autumn leaf, the aging plankton and the dying elephant into life-giving nutrients. The remains of decomposition, humus, fertilizes prairies and jungles. The floor of a redwood forest is literally soundproofed by the accumulated humus of fallen redwoods. Above, a mastlike creaking can be heard. The "masts" are the living redwoods that rise from the humus into the path of sea winds blowing off the Pacific.

Through such decomposition, death is as important as sex in perpetuating life. In this sense, there is no such thing as "waste" in the natural environment. The environment is prepared to dilute, degrade and recycle byproducts of energy and materials dissipation into life processes. The environment is thus dependent on and structured for waste-receiving.

For most of his life, man has lived in harmony and benefited from this ecological ingenuity. The vultures and the bacteria have been equal to the feces and the garbage, the carcasses and the corpses, of a hunter's civilization. Energy conversion at its height was often a matter of throwing another log on the fire. The nomadic nature of the hunter served to disperse his modest wasteload.

As man began to expand his expectations and his range of activities, anti-ecological forces were set in motion. He learned to resist disease and adapt to different environments until he became the most numerous of all large animals and the only worldwide species in existence. Natural controls on man's population size and distribution, which also served to control his waste-making, have been surmounted. In their lifetime, our children may live in a world populated by 15 billion. And man's potential distribution now embraces another "planet," the moon, whose waste-receiving capacity is not yet understood.

In his rise to dominance, man has forsaken nomadism for permanent settlements. This trend serves to concentrate the wasteloads of a population-rise out of control. Our children's children may live to see a city with a population of 1 billion.

Living in cities, man has learned to expand his capabilities for energy conversion. Instead of throwing another log on the fire or trapping wind in ship sails, he has learned to generate energy from coal, oil, and the atom. He has learned to exploit not only the organic wealth of the planet but its inorganic wealth: iron, zinc, copper, and aluminum. He has learned to make his own synthetic products: plastics, nylon, rayons, and pesticides. To convert raw materials into products, he has developed machines as numerous and diverse as insects, from sewing machines to nuclear reactors.

This growth in energy and materials conversion generates a wasteload whose diversity, volume, and complexity far exceed a wasteload fueled simply by uncontrolled population growth and urban concentration. Besides receiving sand grains, autumn leaves, and human excrement, our rivers receive dumped spoil from navigation projects, hot water from power plants, pesticide runoff from farms, soap suds from dishwashers, and acid mine drainage from mines. Besides receiving oxygen from plants and water vapor from lakes, our atmosphere receives carbon monoxide, lead aerosols, fluorides, hydrocarbons, sulfur dioxide, and sonic booms from smokestacks, auto exhausts, and jet engines. Besides receiving fallen redwoods, corpses, and prune pits, our land is receiving items that cannot be readily consumed by bacteria or vultures: aluminum cans, junk autos, plastic cups, copper tailings, and DDT residues.

The tons, cubic yards, and decibels of waste being generated by energy and materials conversion is now prodigious by any scale. Each year, the United States disposes of 7 million autos, 20 million tons of waste paper, 25 million pounds of toothpaste tubes, 48 billion cans, and 26 billion bottles and jars. Our waterways receive some 50 trillion gallons of hot water plus unknown millions of gallons of chemical and organic wastes from factories, canneries, farms and cities.

Yet this wasteload, however large, promises to grow at a rate faster than in the past. From throwaway gum wrappers and paper plates, the "disposable" industry has expanded into paper ties, dresses, bedsheets, and jewelry. The pressure to cut maintenance and labor costs and keep abreast of changing fashion propels the switch from reusable to one-use items. This "one-use" affluence adds a new dimension to wasteloads. The average American family uses 850 cans yearly. In one year, for example, consumers of Col. Sanders' Kentucky fried chicken disposed of 22 million foam polystyrene containers, 31 million paperboard buckets, and 110 million regular dinner boxes. Each year, the babies in the United States have their diapers changed some 15.6 billion times. The diapers, being cloth, are generally washed and reused. Now three large corporations are [producing and advertising] disposable diapers. Large machines turn wood pulp into 300 disposable diapers per minute. Almost as an afterthought, one corporation has set up a Flushability Task Force to determine just how successfully disposable diapers can be flushed down the toilet.

Through such advances as disposable diapers, each American now produces about 5 pounds of trash each day. The *pro-*

rated amount of solid wastes from industrial and chemical processes raises this figure to 10 pounds per person, according to *Cleaning Our Environment,* a 1969 report by the American Chemical Society. Comedian Milt Kamen recently told how his neighbors became suspicious of him because, as a bachelor non-cook, he had no garbage to put out his front door. Kamen says, "I went to the supermarket and I bought some bachelor garbage—frozen garbage with no defrosting. You just toss it out—chicken bones, coffee grounds, egg shells. I'd get garbage delight—lobster shells, an empty champagne bottle. This impressed my neighbors. They called me 'Mr. Kamen' and they smiled at me and said, 'Drop in to see us, Mr. Kamen—and bring your garbage!' "

Animal wastes, chemical fertilizers, and pesticides now make rural as well as urban areas prime sources of waste. The same carefree attitude toward waste-making pertains. The massive application of insecticides by aerial spraying prompted Dr. Kenneth Boulding to remark, "The response of the insecticide industry to insects is like the response of the United States Air Force to Communists."

Loose mines and rusting barbed wire once characterized the leftover debris of war-making; fallout from nuclear bombs and chemical and biological warfare adds a new dimension to war's waste-making potential.

Traditionally, broad-scale energy and materials conversion has been practiced by a minority of the world's population; the United States uses about 50 percent of the raw-material resources consumed in the world. Our 6 percent of the world's population produces 70 percent of the world's solid wastes. The ambitions of developing nations promise to make intensive energy and materials conversion and its corresponding wasteloads global in nature. Yet this planet is being severely stressed by the wasteloads of just one United States, much les 10, 20, 30 or 100 United Stateses. We often tend to think that having the entire world's 3 billion people live American-style would be heaven on earth, a moral goal worthy of all faiths and persuasions. But from an ecological viewpoint, the wholesale use of DDT in Asia, the universal acceptance of flushable diapers, a world wide chain of carry-out chicken stands, two cars and a camper trailer in everybody's garage, and the advent of one-use, nonreturnable, nondegradable homes could be disastrous.

The uncontrolled growth of population, cities, and industry creates an ugly anomaly. With the help of Mr. Clean and other commerical folk heroes, we exhibit an unrivaled concern with hygiene. We probably shampoo our hair, deodorize our armpits, and shave our faces with more regularity than any civilization to date. We have, naturally enough, begun to mechanize hygiene: battery-run toothbrushes, electric shavers, washing machines, and dishwashers. Yet aboriginal hunters must rank as cleaner creatures than we, simply because they were unable to dirty their living spaces to the degree we can. While we deodorize our ampits, our wasteloads overcome the waste-receiving capacity of the natural environment in a rather indiscriminate form of chemical and biological warfare.

11

THE AIR OF POVERTY

PAUL P. CRAIG AND EDWARD BERLIN

A penalty we pay for living in a technological society is exposure to many chemicals and elements from which we are protected in a natural environment. The effect of exposure to low concentrations of pollutants often tends to be epidemiological in character—that is, large segments of the population may be affected; but it is difficult, or impossible, to point to a particular individual who has been demonstrably injured. A classic example of this is found in the well-known studies linking smoking and cancer. Even now, when scientific evidence linking the two is overwhelming, there remain many who are unconvinced by the lack of unambiguously identifiable, individual victims.

As a result of man's industrial activity, he has succeeded in redistributing lead originally in the earth's crust in such a way as to bypass nature's biological protective mechanisms, and thereby poison himself. In this article we summarize the effects of man's activity and indicate some of the types of studies and actions which are essential if lead poisoning is to be understood and eliminated.

Lead is a toxic element with no known beneficial function in human metabolism. Nevertheless, today lead levels in the blood of average Americans exceed one-fourth of those considered diagnostic for classical lead poisoning in adults. This is largely a result of industrial activity. Total body content of lead in Americans increases with age and is greater than that of inhabitants of other countries. Lead can produce liver, kidney, and brain damage, and deterioration of the central nervous and reproductive systems. Children are especially susceptible to the effects of lead poisoning, including mental retardation and other signs of central nervous system involvement.

The detrimental effects of lead have been recognized in this country and abroad. Chronic exposure to lead at levels typical of urban environments is known to produce biochemical changes in healthy adults. In animals, injury has been observed at lead

From *Environment, 13*:2–9 (June, 1971). Reprinted by permission of The Committee for Environmental Information.

exposure levels typical of those experienced by urban Americans.

The chronic effects of sub-toxic body burdens of lead in city dwellers involve reduction in the activity of enzymes essential for hemoglobin (a protein pigment in red blood cells) synthesis. Lead can also destroy existing red blood cells, at least at toxic body levels. There is grave concern about whether these detrimental effects of lead can cause important physiological deficits (even when present at "normal" body concentrations) in children and in people who have inherited defects of red blood cell structure or function, or defective ability to manufacture hemoglobin. There is a critical need for intensive epidemiological studies of people in these categories, such as inner city children with dietary deficiencies which exacerbate the effects of exposure to lead. In the meantime, we should take steps to reduce man's intake of this poison.

In what follows, the effects of lead on biological systems will be discussed. Because all the known effects are deleterious, because there is no known beneficial effect of lead in man, in other animals, or in plants, and because the concentrations of lead in ambient (out-of-doors) air continue to increase, we urge that steps be taken to restrict any further accumulation of lead in the atmosphere.

EFFECTS ON HUMAN BEINGS

Lead poisoning is one of man's earliest self-inflicted diseases. The extent of the problem has been recognized in a report prepared by R. S. Morse under the sponsorship of the Department of Commerce which states:

> Lead has been known to be toxic for over two thousand years and, in spite of its recognition as an industrial poison, it continues to be the cause of numerous outbreaks of chemical intoxication of industrial or accidental origin . . . lead is so widely used in modern technology that occupational health and public health authorities must always be alert to controlling the hazard.[1]

During the early years of the twentieth century lead was widely used in paint. In 1918 it was estimated that 40 percent of all painters showed evidence of lead poisoning. Today, obvious symptoms of lead poisoning are most frequently found in ghetto children who ingest peeling paint and putty. Ghetto children are also exposed to high concentrations of atmospheric lead from automotive emissions.

Between 1954 and 1967, 2,038 children were treated for lead poisoning in New York City, of whom 6.3 percent (128 children) died. It has been estimated that abnormally high levels of lead in the blood occur in as many as 225,000 children.[2-4] According to H. L. Hardy, lead produces

> . . . a variety of changes in practically all portions of the nervous system, some of which are unquestionably harmful. The effect of acute lead poisoning on the brain is extreme. Central nervous system damage caused by lead is marked by destruction of various types of brain cells and degeneration of capillaries and blood vessels throughout the entire structure of the brain.
>
> Lead encephalopathy [disease of the brain] can cause brain hemorrhage, accumulation of fluid, swelling or shrinking of the brain and atrophy of the convolutions. There are serious and widespread disturbances of blood circulation throughout the brain in acute cases. Lead damage to the brain can cause convulsions, delirium, or coma. It can result in severe headaches, blindness, paralysis, mental retardation, or death.[5]

J. R. Goldsmith reported, "inorganic lead in sufficient amounts is implicated as a causative agent in . . . liver and kidney damage.[6] The most common effect on the kidney seems to be a deterioration of the arteries of the kidney, which in time can produce a crippling atrophy (wasting away) of the organ. It is not clear, however, how great a part other, nonlead-related causes may play in this damage.

Lead toxicity is implicated in reproductive failure. Alice Hamilton reported on lead-induced injury in both male and female germ (reproductive) cells:[7] A. Cantarow and M. Trumper say, "there can be little doubt that exposure of mothers to lead has a damaging effect upon fertility, the course of pregnancy, and the development of the fetus."[8]

LOW LEVELS OF LEAD

Low-level exposure to lead is known to produce biochemical changes in man . . . It is well-documented that lead toxicity has a substantial inhibiting effect on red blood cell development. The Morse report recognized this fact. The low-level exposure hemoglobin problem was summarized by the California Department of Public Health:

> There is evidence that exposure to moderately low lead levels may produce abnormalities in the synthesis of porphyrins (substances necessary

for the production of hemoglobin and other compounds in the human body). With moderately high occupational exposures, a considerable increase in delta-amino-levulinic acid (a substance from which porphyrins are formed by the body) in blood and urine can be detected prior to other clinical or biochemical symptoms. With a loss in capacity of reticulocytes (young red blood cells) to synthesize porphyrins, hemoglobin concentration is altered (thus, perhaps, the transport of oxygen and carbon dioxide). Survival time of red blood cells is also lessened. Some decrease in other porphyrin structures, cytochromes, myoglobin, peroxidase and catalase may be expected. The most specific, but not the only, site of damage by lead in porphyrin biosynthesis is the inhibition of amino-levulinic acid dehydrase[9]. . . .

Studies of animals have demonstrated toxicity associated with chronic exposure to lead at levels similar to those experienced by man.

CHILDREN SUSCEPTIBLE

Children are especially susceptible to lead poisoning. The California Department of Public Health reported that

> Children are much more susceptible to lead intoxication than adults. Encephalopathy and mental deterioration in lead-poisoned children have been well documented. One study disclosed that 200 small children had blood lead levels of 0.014 to 0.030 milligram per 100 grams (approximately 14 to 30 micrograms lead per 100 milliliters blood) while 100 mentally defective children showed 0.04 to 0.08 per 100 grams of blood (40 to 80 micrograms lead per 100 milliliters blood). Aminolevulinic acid levels in the blood of these latter children were also high. It has been stated that an upper limit for blood lead in children should be 0.04 milligram per 100 grams (40 micrograms lead per 100 milliliters blood). *This figure already borders on the lower value found in affected children, though general population studies of children have not been done.* . . (emphasis added).

Dr. Joseph Cimino, medical director of New York City's Poison Control Center, has concluded that chronic subclinical (no overt symptoms) level lead poisoning is responsible for the underachievement of children in many areas.

Delayed physiological effects have been observed in children poisoned by lead. In febrile (feverish) diseases and in disturbances of calcium-phosphorus metabolism, lead may be mobilized from the skeletal system (the major long-term lead reservoir) and cause recurrences many years later. A classic case of this sort occurred in Australia. Water from veranda roofs painted with lead-containing paint was consumed by children with no immediate ill effect. These children suffered renal disease, gout, and uremia at 25 to 30 years of age.

Levels of lead in the blood of children from low income areas in Chicago are substantially above those found in adult populations. Of 68,000 children tested during 1967-1968, 5.8 percent had blood lead levels of 50 micrograms lead per 100 milliliters of blood or higher. This latter group of children was referred for treatment for lead poisoning.

OTHER SUSCEPTIBLE GROUPS

We have in our population large numbers of people who have inherited defects in the structure and function of the hemoglobin molecule or in the enzyme system that protects the red blood cell against premature death. These people form identifiable groups who are at increased risk with exposure to lead. In the hereditary disease called thalassemia, there is impairment of synthesis of the protein component of hemoglobin. This leads to formation of red blood cells deficient in hemogoblin, cells that are destroyed by the body at an increased rate. People with this disease are particularly sensitive to lead, either because it interferes with hemoglobin synthesis or because lead contributes to hemolysis (liberation of hemoglobin from red blood cells), thus causing early destruction of red blood cells. This defect is widely distributed among peoples of Mediterranean origin.

In a French clinic two patients exposed to lead showed classical signs and symptoms of lead poisoning. Both of these patients were natives of Senegal, employed at the time in a battery-making factory. One of them had a blood lead concentration of 15 micrograms lead per 100 milliliters of blood when he came to the attention of the clinic. His anemia was due, in addition to lead, to sickle-cell anemia, another inherited defect in hemoglobin formation. The other patient, with thalassemia, had a relatively low level of lead in the urine, a similar degree of anemia, and no recording of his blood lead level. The authors who studied the cases called attention to an association, reported before, between thalassemia and chronic lead poisoning. The patients responded to treatment for lead poisoning.

Professor Sumner Kalman, of Stanford

University, has observed that there are many diseases of hereditary origin where abnormal hemoglobins are formed, and where the patient is susceptible to hemolytic anemia. Far too little research has been done on these problems. This whole gamut of illnesses should be carefully explored to ascertain the role of lead and other environmental insults in red blood cell survival. . . .

There is a wide range of kidney function in the normal population. Kidneys may be normal even though they regularly excrete drugs and chemicals at a rate 40 percent below the mean. There also are a large number of people with kidney insufficiency. Among them, excretory rate for almost any substance extends all the way to zero. Because lead in the blood is excreted primarily by way of the kidneys, these people may face greater risk from lead poisoning.

LEAD ADDITIVES IN GASOLINE

The lead burdens carried by man today are in large part due to his use of lead. Geochemical evidence has shown that in a natural environment human lead levels may well have been a factor of 100 below current levels. Human activities have introduced lead into areas as remote as the Greenland ice sheet. In Greenland, lead levels rose from below 0.001 parts per million (ppm) of snow in the pre-Christian era to more than 0.2 ppm of snow today. Almost all of the increase in lead concentration occurred within the last 50 years. The sharp rise in lead level corresponds closely with the introduction of tetraethyl lead into gasoline in 1923. Today lead is used as an additive in almost all automotive gasoline made in the United States. The amounts added range from two to four grams per gallon. In the United States about 300,000 tons of lead were used in gasoline additives in 1968. This amounts to about 25 percent of the total lead used in the United States. In 1966, motor vehicles discharged 190,000 tons of lead into the atmosphere.[1] Recently, it was estimated that the removal of lead from automotive fuel would "eliminate the 500 million pounds of lead which is currently being emitted from the exhaust pipes of the nation's automotive fleet."[10]

The average lead content of the earth's crust is about 10 to 15 ppm. Lead content in city dust ranges up to about 1 percent by weight [10,000 ppm]—near that of many lead ores. Concentration of lead in surface soil at roadsides is usually many times higher than that at greater distances from traffic. Correspondingly, plants grown near roads are often ten or more times richer in lead . . . than those grown farther away. Most of this lead is on the surface of the plant. Little lead is taken up by the roots of flowering plants since, once in the soil, lead is bound tightly in insoluble lead compounds. On plants along highways lead levels as high as 3,000 ppm have been recorded, the levels often exceed 100 ppm.

Careful washing of plant parts collected from vegetation near highways served to reduce the lead content to levels about equal to that observed in the same type of plant parts collected some distance from the highway. The effectiveness of washing techniques in removing lead from plants was related to the nature of the plant's surface. Where the surface was rigid and smooth, mild washing procedures tended to remove high percentages of lead. In contrast, more vigorous washing techniques were required to remove high percentages of lead from flexible, rough, hairy surfaces. Obviously, forage crops will not be subjected to rinsing or careful washing, and domestic animals could eat the increased lead deposits.

Levels of lead in urban air range from about 1 to 10 micrograms per cubic meter, but atmospheric measurements taken near automobile traffic may extend to above 40 micrograms per cubic meter, depending on proximity to traffic and traffic density. A concentration of lead in the atmosphere in Los Angeles County as high as 71.3 micrograms per cubic meter has been recorded during a peak traffic period. In mid-Manhattan, daily averages of 7.5 micrograms per cubic meter are found. In San Diego the average atmospheric lead level is increasing by 5 percent per year, and week-long averages of 8 micrograms per cubic meter now occur. This contrasts with the 1.5 micrograms per cubic meter limit adopted by the California Air Resources Board and the 0.7 microgram per cubic meter legal limit set by the Soviet Union. Urban atmospheric lead levels and lead in rainfall are correlated with local gasoline sales. Lead in rainfall in the U.S. averages about 0.034 ppm with peak concentrations as high as 0.3 ppm compared to a legal limit in drinking water of 0.05 ppm. Water supplies actually contain less lead (0.017 ppm) than rainwater, probably because some of the lead precipitates as insoluble salts. Lead flow to the oceans is probably about 40 times the primitive level, resulting in surface lead concentrations about 5 times greater than in underlying layers of water.

About 95 percent of ingested lead—that is, lead taken in from food and beverages—is quickly excreted. Most of the 5 percent which remains is absorbed into the blood and later excreted in the urine. A small portion of this absorbed lead, however, is retained in body tissues, mostly the bones, so that over a period of years a body burden of lead accumulates which varies with changing conditions of lead exposure. Lead in the bones is insoluble and presumed to be inert. Under certain conditions, however, the body can release deposited bone lead into the blood stream. In children, the mobilization of bone lead is unpredictable and may cause lead poisoning even though lead is not being eaten.

Whereas only 5 percent of lead intake from food and water is absorbed—even temporarily—in the blood and tissues, 30 to 40 percent of lead inhaled into the lungs is absorbed. Because of the difference in absorption rates, the amount of lead absorbed via the lungs may be similar in magnitude to that absorbed through the intestinal tract, even though for most people the total quantity of lead eaten with food and water is usually several times the total inhaled in air.[6]

The contribution of inhaled lead to the body burden, however, also depends on exposure. In a Los Angeles study, blood lead levels in persons living near a freeway were markedly higher than levels in persons living a mile from the freeway in relatively clean coastal air.[6] This means that lead in the air may be the largest contributor to absorbed lead in some persons living in areas exposed to heavy auto exhaust pollution.

Lead emitted from automobile exhausts is particularly suited for retention in the atmosphere and eventual entrance into the body. About 75 percent of particulate lead from gasoline combustion is less than 0.90 micron (less than 0.0000351 inch) in mean diameter, a size that easily reaches the alveoli (air cells of the lungs).

The most common test for measuring the

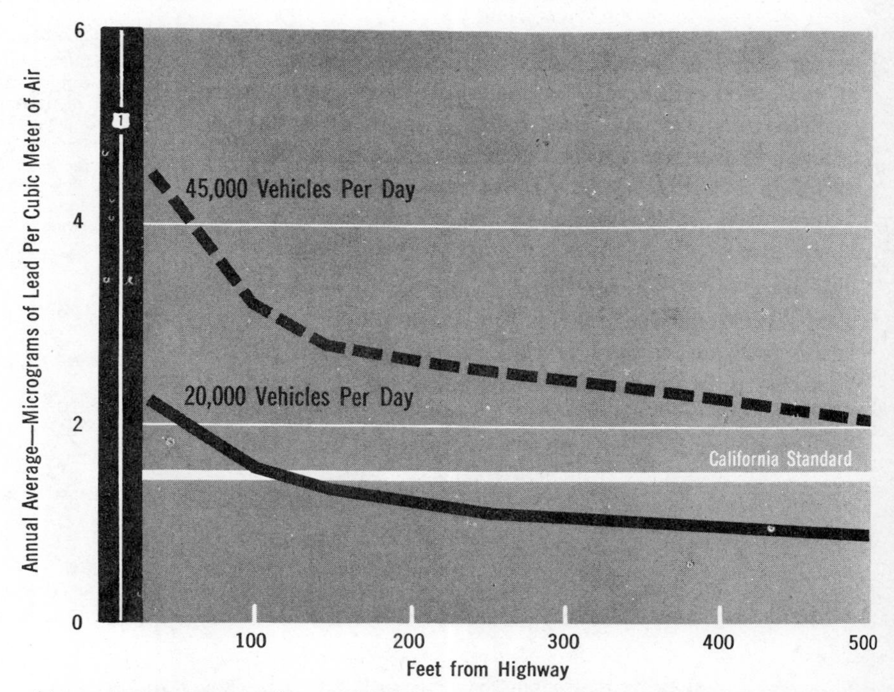

Atmospheric concentrations of lead from vehicles vary according to traffic volume and distance from the highway, as can be seen in this figure which illustrates the situation from mid-1967 through mid-1969 along different sections of U.S. Highway 1 in New Jersey, supporting 20,000 and 45,000 vehicles a day. . . . The lead concentrations [are compared with a] recently enacted California Standard . . . (Adapted from *Environmental Science and Technology,* April 1970, p. 320.)

body lead burden measures the blood lead concentration. In adults, the industrial guideline for lead poisoning is 80 micrograms of lead per 100 milliliters of blood. The mean blood lead level in the United States is about 20 micrograms lead per 100 milliliters of blood, with higher levels demonstrated in persons who work or live close to automotive traffic and lower levels for people who reside in rural areas.

However, the blood concentration is not the most sensitive index of the total body burden of lead. H. A. Schroeder and I. H. Tipton have shown that lead concentrations in the tissues of Americans are generally much higher than those of Europeans, Africans, Asians, and Mideasterners. Furthermore, these body burdens of lead increased sharply with age in the case of the American subjects, but not in the foreign subjects. The authors stated:

> It is likely that atmospheric lead from motor vehicle exhaust largely accounts for increased exposures [of American subjects], and that inspired [inhaled] lead may make up a sizeable portion of the total amount absorbed by the body. . . . In view of the steadily increasing annual pollution of air and soils with lead from motor vehicle exhausts, innate toxicity in exposed human beings may appear.[11]

Probably one-third or more of the lead in city-dwelling Americans is attributable to inhalation of airborne lead from automotive emissions. Additionally, an appreciable portion of the lead in food may originate from fallout of lead introduced into the atmosphere by automobiles. Removal of this source of lead would lower body lead levels, thereby lowering the extent of the damage now being inflicted upon children, as well as upon other members of our society who may be particularly sensitive to this entirely unnecessary environmental stress.

DECISIVE ACTION REQUIRED

In his 1965 address before the Section of Occupational Medicine of Britain's Royal Society of Medicine, Sir Austin Bradford Hill stated:

> All scientific work is incomplete—whether it be observational or experimental. All scientific work is liable to be upset or modified by advancing knowledge. That does not confer upon us a freedom to ignore the knowledge we already have, or to postpone the action that it appears to demand at a given time.[12]

At best, knowledge of the scientific and medical realities of atmospheric lead pollution is incomplete. But existing knowledge demands decisive, expeditious action, for it compels but one conclusion: Atmospheric lead, at levels now common in urban areas, is a human health hazard.

This paper is based upon legal documents used in action by the Environmental Defense Fund before the Department of Health, Education and Welfare; Environmental Protection Agency; and the California Air Resources Board. Contributions in the form of comments, criticisms, or affidavits from many scientists have contributed to this work. We particularly acknowledge Drs. Harriet Hardy, Henry Schroeder, Clair C. Patterson, T. J. Chow, Sumner Kalman, Russell Hewlett, Joseph Cimino, and Robert Debs. [Many of the footnotes and references in the original paper are omitted here (Ed.)]

REFERENCES

1. Morse, R. S.:*The Automobile and Air Pollution, a Program for Progress,* Part II, U. S. Department of Commerce, Washington D. C., 1967.
2. Oberle, M. W.: Lead Poisoning: A Preventable Childhood Disease of the Slums, *Science 165:* 991, 1969.
3. Elwyn, D.: Childhood Lead Poisoning. *Scientist and Citizen 10(3):* 55, 1968.
4. English, M.: In *Look Magazine*, October 21, 1969.
5. Hardy, H. L.: Lead and Health. *Environment 10:* 80, 1968.
6. Goldsmith, J. R.: Epidemiological Bases for Possible Air Quality Criteria for Lead. *Journal of the Air Pollution Control Association 19:* 714, 1969.
7. Hamilton, A.: *Industrial Toxicology*. Hoeber, New York, 1934, pp. 43–44.
8. Cantarow, A., and Trumper, M.: *Lead Poisoning.* Williams and Wilkins, New York, 1944.
9. *Lead in the Environment and its Effects on Humans.* California Department of Public Health, Berkeley, 1965.
10. Glogan, R. C., senior vice-president of Engelhand Minerals and Chemicals Corp: Testimony given before the Senate Committee on Public Works, March 25, 1970.
11. Schroeder, H. A., and Tipton, I. H.: The Human Body Burden of Lead. *Archives of Environmental Health 17:* 965, 1968.
12. Hill, A. B.: The Environment and Disease: Association or Causation. President's address before Section of Occupational Medicine, Royal Society of Medicine. *Proceedings of Royal Society of Medicine, 58:* 299, 1965.

12
MERCURY IN MAN

NEVILLE GRANT

Alamogordo, New Mexico is not a large town. Located in the sparsely populated southern portion of the state, it has few distinctions. However, in a sense, Alamogordo is symbolic of two ongoing threats to the environment that first gained national recognition there. In 1945 the first atomic weapons were exploded in nearby White Sands. Nuclear radiation as an environmental hazard became a reality. Twenty-five years later, in late 1969, an Alamogordo family ate mercury-contaminated pork and this country awoke dramatically to the possibility of widespread mercury pollution. The paths of these two environmental contaminants—radiation and mercury—converge at the genetic material of living cells. The disruptive effects of mercury on cells—called radiomimetic because they are similar to those caused by radiation—are greater than the effects of any other known chemical. Do present mercury levels in the environment pose a threat to man? It would seem that the answer is yes. The magnitude of this threat is as uncertain as that of low levels of ionizing radiation. . . .

In the early 1950s the Minamata Bay disaster alerted the Japanese to the sequence: mercury discharge from industrial waste into water → concentration in fish → disease in people. The lesson, however, was not sufficient. In 1964 the Agano River in Niigata Prefecture was contaminated by mercury in industrial effluent: 26 persons fell ill; 5 died. Elsewhere reports appeared of poisoning in large groups of people following the ingestion of mercury-treated seed baked into food. These incidents invariably occurred in underdeveloped regions where malnourishment was prevalent. Sweden critically examined the extent of mercury contamination in fish and birds in the mid-1960s, took steps to eliminate mercury as an environmental pollutant, and set guidelines to avert possible human poisoning. Reports began to appear in late 1969 of elevated levels of mercury in partridge and pheasant in various parts of Canada and the U.S. The stage was set for Alamogordo.

From *Environment, 13:*2–15 (May, 1971). Reprinted by Permission of The Committee for Environmental Information.

In September 1969 Ernest Huckleby obtained floor-sweepings from a granary—millet seed treated with a methyl mercury fungicide. It has been an established practice that all treated seed is mixed with a red coloring to avoid what happened next. Huckleby fed the seed to seventeen hogs he was raising. Subsequently, several hogs became ill. He chose one that appeared to be well, butchered and froze it, and began feeding it to his family; the other hogs were sold. Several weeks later three of the Huckleby children fell ill. The diagnosis, when finally established, was acute mercury poisoning. Side meat from the hog contained 27 parts per million (ppm) of mercury. The three children suffered irreparable neurological damage and a child, born of the then pregnant but asymptomatic mother, had convulsions after birth—a common manifestation of mercury toxicity in the newborn.

Thus began a hard and still unresolved battle over the use of organomercury compounds as fungicides in agriculture, a battle that Sweden had decided some years earlier in favor of suspension. . . . One thing seems clear: the Alamogordo incident was not unique. At the present a reappraisal of mercury compounds of all types used in agriculture is underway. Sweden has shown that more stringent rules of application of the less toxic and more easily degradable mercury compounds can markedly reduce food contamination without a significant reduction in crop yield.

In March 1970 fish in Lake St. Clair on the U.S.-Canadian border, and subsequently in many lakes and rivers in the U.S.[1] and Canada, were found to be contaminated with mercury at levels (1 to 10 ppm) comparable to those implicated in the Niigata disaster. In most, but not all instances, the source of pollution could be traced to industrial effluents from chemical plants, pulp factories, and the like. Fishing industries collapsed in certain areas as news spread of high mercury levels. Economic disaster visited many whose livelihood was sport and commercial fishing. The state of Alabama prohibited commercial fishing in 51,000 miles of waterways and sought federal disaster relief. They were turned down because the situation was not the result of natural causes. The Canadian government, however, has offered low-interest loans to those fishermen who were affected by high mercury levels. The Food and Drug Administration—understaffed and unprepared—began a widespread survey of all foods. Agencies both public and private began testing programs. Symposia and congressional hearings were held. Then ocean-going tuna and swordfish were found high in mercury—the source as yet uncertain. The final injustice: some health food tablets, made from the compressed liver of seals caught in the Pacific, were loaded with mercury (36 ppm).

This stacatto series of events was difficult to digest, because full information was not available on which to base an evaluation of the seriousness of the situation. Many of the pieces are still missing, but the picture is beginning to emerge. In order to evaluate new information as it appears and to formulate policy decisions, attention must be given to several broad points:

Where does mercury that contaminates the environment come from; how is it dispersed; and in what form does it reach animals and man?

How are the different forms of mercury handled by the body?

At what level of exposure does mercury pose a problem for humans, and how close are we to those levels?

What is the extent of present mercury contamination? What is an appropriate maximum permissible level of mercury contamination in food?

MERCURY IN THE ECOSYSTEM

Where does mercury that contaminates the environment come from; how is it dispersed; and in what form does it reach animals and man?

Mercury is present in all terrestrial soil and rock. The average amount is approximately 0.05 ppm. Certain large mercuriferous belts which occur in sediment and volcanic rock consisting largely of cinnabar (HgS) may contain from 1 to 30 ppm. During the weathering of rocks and deposits mercury may enter the sediment of streams, lakes, and oceans.

Perhaps the largest quantities of mercury added to the environment by man are released with the burning of natural fuels. From available evidence it appears that U.S. coal may average between 0.5 and 3.3 ppm mercury; at present rates of coal consumption, between 275 and 1,800 tons of mercury are probably being released to the air annually.[2] Petroleum may contain from 1.9 to 21 ppm mercury; it is not known how much is released in refining and burning. Bitumens and asphalt contain 2.0 to 900 ppm mercury. The combined use of all these materials may add more mercury to the environment than intentional uses of mercury in industry and agriculture.

The extraction and use of mercury itself is a substantial source of pollution. According to one estimate, mining, smelting, and refining operations released 170,000 pounds of mercury into the environment in 1965[3] About one-third of the annual U.S. consumption of mercury is produced by domestic mines; the remainder is either imported or released from federal stockpiles. Total demand for mercury in 1968 was 5,730,400 pounds[1], of which a significant portion may ultimately find its way into the environment.

An analysis of the uses to which this mercury is put was recently performed by a group at Oak Ridge National Laboratory.[1] They reported that 26 percent of the total mercury demand in 1968 was required for "dissipative" uses, such as mercury fungicides for agriculture and manufacturing, in which mercury is necessarily dispersed to the environment. By far the largest consumer of mercury was the manufacture of chlorine and caustic soda. Twenty-three percent of the total mercury consumption during 1968 was released to the environment as wastes from these plants. Another 23 percent of total consumption was of unknown ultimate disposition; much of this material was fixed in electronic instruments and measuring devices, including thermometers. How and to what extent this mercury reaches the environment is not known. Only 18 percent of the mercury produced was reclaimed for further use, however, and a great percentage of the remainder probably will eventually find its way into the environment in one way or another.

Agricultural uses of mercury, although small compared to the total, are of concern because of their immediacy to man. Mercurial fungicides may enter the food chain of man in a variety of ways: through the spraying of plants; the leeching of fungicide from soil to waterways; the translocation of mercury from seed to plant (crops grown from treated seed may contain twice the mercury content of untreated crops; this level appears to be extremely low and would seem to pose no significant threat); and feeding of treated seed to livestock and poultry. The real and potential threat of ingestion of methyl mercury-treated seed by birds, mammals, and their bird predators, has been well documented.

Most recent concern over mercury has been focused on water pollution and the resultant contamination of fish. The degree to which this situation is man-made is still in dispute. The amount of natural mercury in streams, rivers, and lakes averages 0.03 part per billion (ppb) (range 0.01 ppb to 0.1 ppb). Oceans average 0.03 to 5.0 ppb. Waters near mercury deposits contain 5.0 to greater than 100 ppb. It has not yet been shown what percentage of this total mercury in water is in the form of the water soluble and toxic methyl mercury. Studies are in progress on this point. The conversion of inorganic mercury to organic by microorganisms ("methanogenic bacteria") has been shown[2] and this reaction appears to be ubiquitous. Thus, lakes contaminated by simple inorganic mercury act as a continuous source for the production of methyl mercury. It has been estimated that this mercury may remain in the environment for as much as 100 years.

Methyl mercury is concentrated in fish approximately 5,000-fold that in the surrounding water by entering through the gills during respiration or by ingestion of mercury adsorbed on phytoplankton. Once bound to protein, the half-life (the time required for half the mercury to be eliminated) in fish has been judged to be two to three years and the binding is of such strength that it is not disrupted by freezing, boiling, or frying. Ninety-five to 100 percent of the mercury in fish is methyl mercury. That environmental pollution is the major cause of elevated levels of mercury in fish is shown by the 100-fold increase in mercury in certain species of fish in Lake St. Clair (0.07 to 0.11 ppm in 1935 to 5.0 to 7.0 ppm in 1970). The natural level of mercury in fish is thought to be up to 0.2 ppm; higher concentrations are probably due to mercury contamination of water, except in water which washes over mercuriferous rock. High levels of natural mercury in water may not necessarily concentrate in fish if it is not in a methylated form. However,

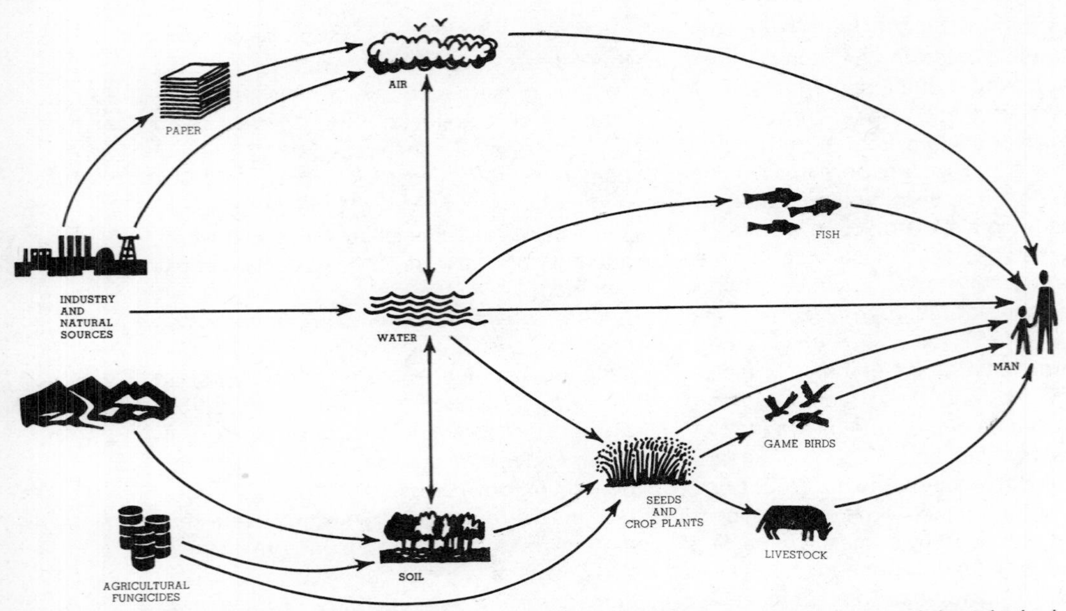

This flow chart shows how mercury reaches man. Although research has now uncovered the principal routes by which mercury reaches man in air, water, and food, the relative importance of the different routes of exposure is still not known. A major uncertainty which remains is the fate and significance of the massive amounts of mercury released to the air by industry, a source of environmental contamination which has just come under investigation.

research is in progress on the possibility that microorganisms on fish gills may be methanogenic, that is, capable of converting inorganic mercury to an organic form. The argument that high levels (greater than 0.5 ppm) in fish are natural overlooks the fact that almost all of this is methyl mercury and therefore potentially dangerous.

MERCURY IN THE BODY

If there is dispute over the sources of mercury in the environment, its effects are still more hotly debated. These effects vary greatly with the chemical compounds of mercury to which the body is exposed.

The two general classes of mercury compounds—organic and inorganic—differ greatly in the extent to which they are absorbed by the body and the degree of damage they may do, once absorbed. As was pointed out in our previous report in May 1969, the most common form of organic mercury found in the environment, and unfortunately the most toxic for man, is methyl mercury. Most mercury compounds can be converted to this form by bacteria once they have been released to the environment; it is this form which will concern us here. Methylation is the process whereby mercury is changed from an inorganic to an organic form by combining with a one-carbon organic molecule. Methyl mercury is far more readily absorbed, and much more slowly excreted, than inorganic mercury compounds. When taken by mouth, for instance, absorption of inorganic mercury is extremely low—less than 2 percent. That some conversion of inorganic to methyl mercury by bacteria may take place in the intestine of some animals has been suggested. Birds fed various non-methylated mercury compounds (as seed-dressing on feed) produce eggs which contain methyl mercury. The extent to which methylation of mercury in the intestinal tract takes place is unknown. Indications are that it may be quite small, but studies on humans, particularly during periods of slow intestinal passage, are entirely lacking. *Intestinal absorption of methyl mercury, however, is close to 90 to 95 percent.*

Once absorbed, methyl mercury is transported by the blood throughout the body, almost entirely by the red cells. It appears to be bound to certain sulfur-containing (sulfhydryl) "binding groups." Thus, measurement of mercury in red cells reflects the level of methyl mercury in whole blood. The relatively slow excretion of methyl mercury may be due to the fact that it is excreted in the bile and then almost totally reabsorbed by the small intestine.

Inorganic mercury, by contrast, is transported approximately 50 percent in plasma and in this form is more readily excreted in urine. Aryl mercury (substances such as phenylmercuric acetate, or PMA—a non-methylated form of organic mercury) is largely converted to the inorganic form and handled as such by the body.

Slow conversion of methyl mercury to the inorganic form, in the liver and other organs, probably also accounts for its delayed excretion. There is some evidence that the liver may contain the capability of methylating mercury, but the extent of this reaction is unknown.

Methyl mercury is excreted largely by the intestine, and evidence indicates that some may be converted to inorganic mercury in the process. Small amounts are excreted in the kidney in both the organic and inorganic form. Thus urinary levels of mercury are a poor indicator of prior methyl mercury exposure.

Mercury compounds have been shown to cause breakage and abnormal chromosome division in concentrations lower than any other known substance—as low as 0.05 ppm. An area of both major concern and incomplete knowledge is how mercury may affect chromosomes—the carriers of genetic material. It is here that the dangers of low-level pollution may be the greatest. Decisions must be made regarding acceptable levels of exposure. Genetic abnormalities occur when chromosomes are damaged or their division disturbed. The derangement of chromosomes by mercury probably depends on its interaction with sulfhydryl groups essential for normal spindle formation that directs an equal division of chromosomes into each newly formed cell. Interference with this segregation will result in some cells receiving unequal numbers of chromosomes (non-disjunction). That this happens readily in plant cells and flies exposed to mercury has clearly been shown. This effect has been shown with aryl as well as alkyl (methyl) mercury. A variety of human genetic abnormalities of unknown cause are the result of non-disjunction. There has been no evidence that mercury has caused chromosome damage to reproductive cells in humans such that genetically malformed offspring have resulted. However, should isolated instances occur, the evidence would be extremely difficult to detect. Recently, evidence of slight but statistically significant increase in chromosome breakage was noted in lymphocyte (white blood cell) cultures from humans with elevated blood mercury levels due to ingestion of contaminated fish. The significance of this observation is not yet clear. Vaginal application of PMA to pregnant mice resulted in fetal death and malformation. The reason for this is not clear, because PMA given intravenously is not detected in the fetus (unborn offspring). despite a large concentration in the placenta (the organ which unites the fetus to the maternal uterus). This is in contrast to methyl mercury, which passes the placental barrier readily. The pertinence of these findings is heightened by the fact that many conveniently available vaginal creams contain PMA. One such product contains 0.02 percent PMA (delivering 16 milligrams of PMA in an 80-gram tube which is designed to be used within one week) and is readily absorbed through the vaginal lining. Its effect on maternal reproductive cell, sperm, and developing fetus is unknown. A furtheir review of this problem is given by M. Shimizu[4]. . .

A most disturbing feature of methyl mercury is its demonstrated ability to penetrate the placental barrier and produce fetal damage. Toxic effects have been noted in the fetuses of experimental animals and in humans at levels that have not produced symptoms in the mother. There is an apparent concentration of methyl mercury in the fetus, as manifested by a higher concentration of mercury in umbilical cord red cells as compared to maternal blood. Developing fetal nerve tissue appears to be more sensitive to the destructive effects. Some evidence suggests that placental and fetal "trapping" of methyl mercury may actually protect the mother. Recently a new factor has received attention. Certain substances such as NTA (nitrilotriacetic acid—introduced as a substitute for phosphates in detergents) have been shown to form chemical complexes with mercury that appear to increase the ability of mercury to cross the placenta and produce damage in the unborn. Wisely, caution has been observed, and NTA has been removed from detergents.

Methyl mercury characteristically penetrates the blood-brain barrier and is distributed in the brain, producing specific symptoms due to the destruction of cells in the cerebellum (balance center), visual and hearing centers, and elsewhere. Inorganic mercury does not readily penetrate to brain tissue except as a vapor.

LEVEL OF EXPOSURE

At what level of exposure does mercury pose a problem for humans? Figures 1 and 2 correlate the relation of estimated onset of tissue damage with concentration of mercury in red blood cells (Figure 1) and brain tissue (Figure 2). In each instance these important assumptions are made: (1) All mercury is in the form of methyl mercury or is converted to it. (2) Exposure is continuous, or nearly so.

What is an Acceptable Daily Intake (ADI) of mercury? Calculation of an ADI (whether for mercury or any other chemical food contaminant) is an extremely difficult task. With regard to mercury, a number of factors must be considered.

The form of mercury must be known. Clearly, the more easily absorbed and more toxic alkyl mercury will require a much lower ADI than other forms of mercury.

Calculations of half-life in the body and distribution (especially to the brain) may vary. Levels observed in humans may be either falsely low (for example, blood collected at varying times after exposure ceased) or falsely high (for example, in autopsy brain tissue concentration in fatal cases) in relation to symptoms.

Damage may occur at levels much lower than can be recorded by present techniques of clinical evaluation. (This is almost certainly true with regard to brain damage and quite possibly true with regard to chromosome damage). Of great interest in this regard is a case reported by Takeuchi of an elderly man (with no symptoms of mercury poisoning) found at autopsy to have brain lesions typical of methyl mercury exposure (from frequent fish consumption).

Variation in individual sensitivity appears to be extremely wide. A safety factor of 100 is generally accepted as adequate when transferring experimental results from animals to man (interspecies variation). A safety factor of ten may be adequate in relating effects in humans to other humans (intraspecies). Both safety factors are generally calculated from a "no-effect" level for the *most* sensitive species.

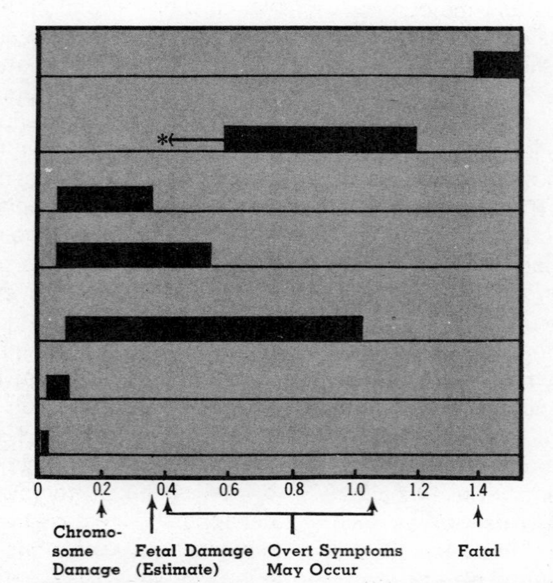

Figure 1. Relation of Methyl Mercury Levels in Blood to Physical Hazard (From Löfroth, G., "A Review of Health Hazards and Side Effects Associated with the Emission of Mercury Compounds into Natural Systems," Swedish Natural Science Research Council, Stockholm, Sweden, 2nd Edition, Sept. 1970 and Berglund, F. et al., "Methyl Mercury in Fish, A Toxicologic-Epidemiologic Evaluation of Risks," a report preprinted from Nordisk Hygienisk Tidskrift, Supplement 4, Stockholm, Sweden, 1971).

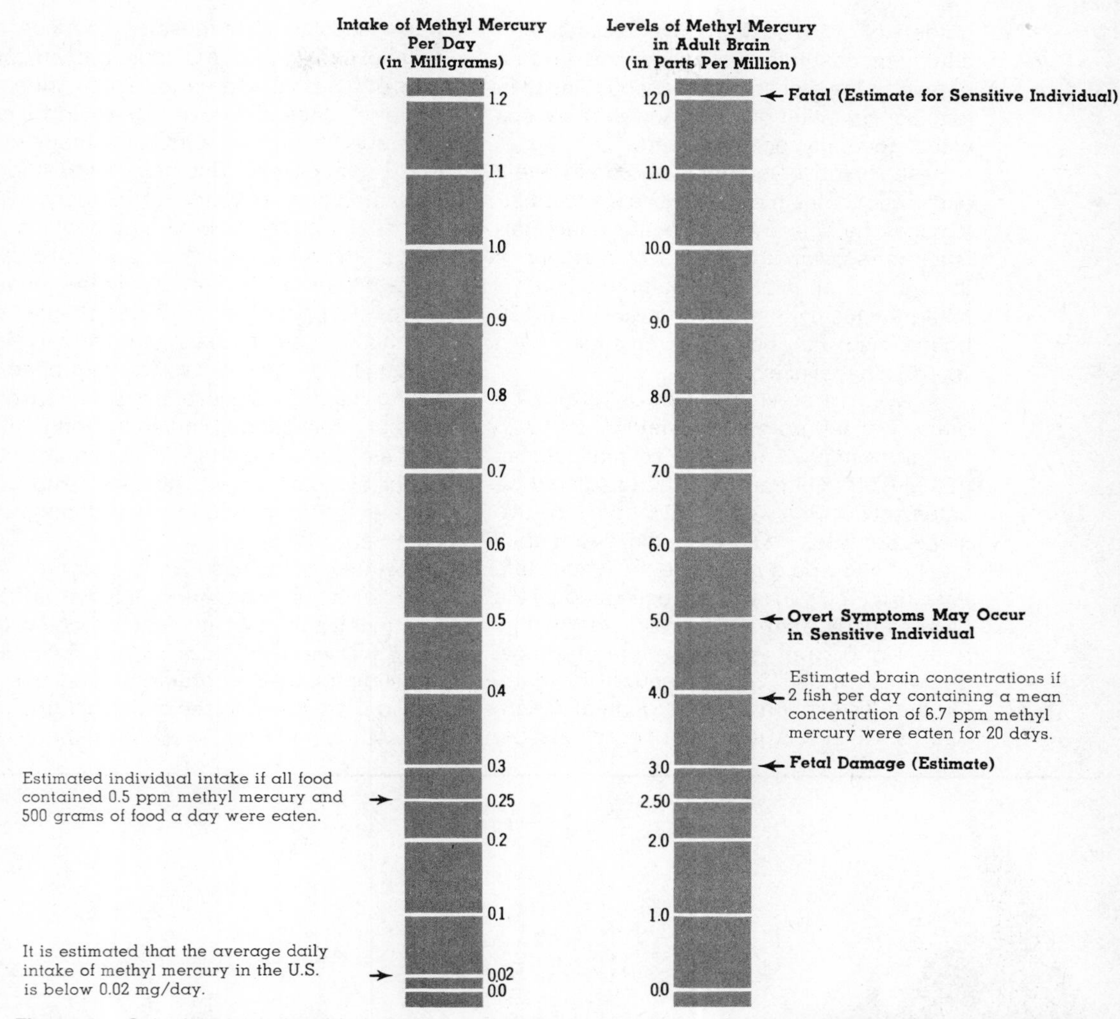

Figure 2. Calculated relationship between methyl mercury intake and levels of methyl mercury in brain tissue. Calculations of brain tissue levels are based upon: (1.) brain distribution of 15 percent of total body methyl mercury. (It has been calculated at 10 to 15 percent.), and (2.) continuous exposure for one year. With an excretion rate of 1 percent per day of total body mercury the indicated level will almost be reached. (From Löfroth, G., "A Review of Health Hazards and Side Effects Associated with the Emission of Mercury Compounds into Natural Systems," Swedish Natural Science Research Council, Stockholm, Sweden, 2nd Edition, Sept. 1970; Berglund, F. et al., "Methyl Mercury in Fish, A Toxicologic-Epidemiologic Evaluation of Risks," a report preprinted from Nordisk Hygienisk Tidskrift, Supplement 4, Stockholm, Sweden, 1971; and Takeuchi, T., "Biological Reactions and Pathological Changes of Human Beings and Animals under the Condition of Organic Mercury Contamination," International Conference on Environmental Mercury Contamination, Ann Arbor, Michigan, 1970).

Considering the many areas of uncertainty to be taken into account, it is not surprising that standards for mercury levels in the diet recommended by various groups have differed widely. One of the most widely quoted "standards," however, is not really a standard at all. In 1967 the World Health Organization published a report in which it estimated a "practical residue limit" in food. This was defined as the concentration of mercury expected in the diet from natural background and natural environmental contamination, from 0.02 to 0.05 ppm mercury in its organic forms. Rather than a standard, this is the irreducible minimum of exposure to mercury that we can expect, and WHO and other organizations have recommended that all efforts be made not to exceed this minimum exposure.

The U.S. Food and Drug Administration, largely in response to concern over contamination of fish, has established an "interim guideline" of 0.5 ppm. This is predicated on a fairly modest intake of fish, as this level, if present in the diet as a whole, would result in total consumption of mercury well in excess of that generally recommended.

Based on observed blood and brain levels of mercury of the most sensitive symptomatic cases, half-life and distribution characteristics, and estimated intake of mercury in toxic cases, an intake of not more than 0.03 milligram methyl mercury per day (for a 150 pound man) has been recommended by a Swedish panel of experts.

An average serving of fish of 150 to 200 grams with 1.0 ppm would yield 0.15 to 0.20 milligram methyl mercury. To maintain an ADI of 0.03 milligram per day this could be eaten only once a week. At 0.5 ppm only two meals per week could be eaten, and a daily meal of fish would reach the maximum safe level of mercury in red cells estimated by the Swedes (0.04 ppm). For unrestricted eating, a level of 0.2 ppm would be the maximum permitted by this recommended standard. On the assumption that the general American fish intake is less than two meals per week (and for other reasons), the FDA has set a level of 0.5 ppm. Clearly, with an upper limit of 0.5 ppm and careful monitoring, the average concentration in fish would be considerably lower—perhaps in the range of 0.2 to 0.3 ppm or less. This is in contrast to the estimated concentration of mercury in the fish at Niigata producing mercurialism—50 percent less than 1.0 ppm and 50 percent greater than 1.0 ppm, mostly in the range of 1.0 to 10.0 ppm. This range has clearly been reached in certain lakes and rivers in North America. *The only difference has been the eating habits of the two areas.* No cases of mercury poisoning from fish in North America have been reported. It is possible that minor symptoms have occurred and gone undetected or that subclinical damage has occurred.

Mercury exposure occurs from other sources in the environment. It is not yet clear how much and in what form this occurs. *If all food averaged 1.0 ppm methyl mercury,* a consumption of 500 grams of food per day would produce an intake of 0.5 milligram per day. At the end of one year, equilibrium at 5

"The devil with the food chain. I *like* mercury."

From *Sidney Harris,* American Scientist.

ppm brain tissue would occur, which is clearly a toxic dose. A market basket averaging 0.5 ppm would yield an intake of 0.25 milligram per day, clearly a dangerous dose. A general level of 0.1 ppm would mean a daily intake of 0.05 milligram. 0.06 ppm in all food would give the Swedish group's recommended maximum acceptable intake of 0.03 milligram per day.

What are the current levels of mercury in food? At present, knowledge of mercury levels in food is fragmentary, scattered, and difficult to come by. . . . No significant monitoring for the element in food in the United States was in effect before 1970. A total diet survey in Britain in 1969 showed the level of mercury to be at or below the level of detection by the method used (0.01 ppm).

In Sweden most of the fish investigated from freshwater and coastal areas were in the range of 0.20 to 1.0 ppm with 1 percent greater than 1.0 ppm and the highest concentration being 9.8 ppm. Ocean-going fish generally range from 0.03 to 0.1, with the exception of tuna with a slightly higher range and swordfish frequently found to have greater than 0.5 ppm.

In Canada one survey of a variety of fish sampled gave levels in a range of 0.02 to 0.42 with an average of approximately 0.1 ppm. The range was considerably higher for fish from Lake St. Clair and Lake Erie. Other surveys have shown higher concentrations in areas related to industry and pulp mills.

In the United States many states have reported areas where the content of mercury in fish was in the range of 0.5 to greater than 1.0 ppm. Most fish, however, living in noncontaminated areas contain less than 0.5 ppm. At the present time a widespread survey is taking place on mercury in the foods of the market basket. These reports are not yet complete; several things, however, seem clear: (1) In general, foods contain less than 0.1 and are usually below 0.06 ppm. (2) "Hot spots" may occur along certain chains, for example, seed- and fish-eating birds which *may* accumulate significant levels. Excessive use of mercury fungicides and the feeding of mercury-treated seed to poultry and livestock may increase the possibility of food contamination. A decrease in mercury content of food can be brought about by an increase in the use of non-mercurial fungicides.

In conclusion it can be said that no widespread epidemic of mercury poisoning is underway in the U.S., perhaps in large part because U.S. eating habits do not include large amounts of fish. However, there is clearly the potential for human damage at present levels of mercury pollution. Much new knowledge must be gained before a safe level of exposure can be confidently set. The difficulty, of course, is the uncertainty of levels that might cause chromosome damage. Certainly evidence of chromosome damage to white blood cells at mercury blood levels commonly found in moderate fish consumers in Sweden gives great pause. New information may well alter what is ultimately considered an acceptable daily intake. In the meantime, prudence would dictate that intensive efforts be made to reduce industrial mercury wastes, to find other, less toxic agricultural fungicides, and to pinpoint and reduce all other sources of mercury contamination of the environment.

In addition, the methods of mercury detection must be standardized in such a way that results are reliable and reproducible. Already some controversy has arisen over this point. More information is needed concerning what percentage of the mercury in the environment is methyl mercury. Finally, a program of regular monitoring of food must be maintained.

Editor's Note: Many of the footnotes and references in the original paper are omitted here.

REFERENCES

1. Wallace, R. A., Fulkerson, W., Shults, W. D., and Lyon, W. S.: *Mercury in the Environment, the Human Element.* Oak Ridge National Laboratory, Oak Ridge, January, 1971, p. 5.
2. Wood, J. M., Kennedy, F. S., and Rosen, C. G.: Synthesis of methyl-mercury compounds by extracts of a methanogenic bacterium *Nature* (London) *220*:173, 1968.
3. Lutz, G. A., Gross, S. B., Boatman, J. B., *et al.*: *Design of an Overview System for Evaluating the Public Health Hazards of Chemicals in the Environment.* Battelle Memorial Institute, July, 1967, NTIS Document PB194 398, p. 4.
4. Shimizu, M.: Distribution of intravaginally administered phenyl mercury acetate in the body and its effect on the fetus. 2. Organ distribution and transport to the fetus. *Journal of the Japanese Obstetric and Gynecological Society 22*:1209, 1970.

13

PCB: THE WORLDWIDE POLLUTANT THAT NOBODY NOTICED

KEVIN P. SHEA

From *Environment, 15:*25–28 (November, 1973). Reprinted by permission of The Committee for Environmental Information.

The polystyrene coffee cup you used this morning, plastic liners for baby bottles, frozen-food bags, bread wrappers, and many other products which come into contact with food contain PCBs or polychlorinated biphenyls. While recently issued federal standards restrict these toxic materials to much less than 1 part per million in most foods, PCB concentrations in the food containers just named have been measured at as high as 6.6 parts per million. No one knows whether these PCBs leach out into the milk held by baby-bottle liners or the coffee held in polystyrene cups; this is just the latest in a long series of riddles surrounding the use of PCBs.

Polychlorinated biphenyl compounds had been in use for 36 years before their discovery in 1966[1] in the flesh of fish from the Baltic Sea set in motion an intensive worldwide research effort to determine the extent of their distribution in ecological systems and the possible danger their distribution may pose to human health. Since their belated discovery as important global pollutants, PCBs have chalked up an impressive record of finding their way into human food chains, sometimes by totally unexpected routes, and have been responsible for huge economic losses in the food industry and, at least in one case, for a serious epidemic of human poisoning. Although important steps have been taken to prevent such episodes from recurring, there remain a number of questions to be answered regarding the uses of PCBs, their effects on biological systems, and their routes of entry into the environment.

Oddly, it was the experience gained in the pursuit of information about another group of environmental contaminants, the chlorinated synthetic pesticides (DDT and its relatives), that both helped and hindered in tracking down the various pathways through which PCB might enter and spread throughout ecological systems. By the middle 1950s analytical techniques had been developed (gas chromatography) that were able to detect very small amounts of environmental contaminants. Analytical chemists, however, could not, with any certainty, identify a particular compound unless they knew precisely what they were looking for. In the case of DDT, for example, they could compare known samples of DDT with results obtained from chromatographic analysis of environmental samples.

In 1964 the Swedish Royal Commission on Natural Resources employed Dr. Soren Jensen of the University of Stockholm to begin a study of chlorinated pesticides in the Swedish environment. It was during this study that Dr. Jensen and his colleagues began to see evidence on their chromatograms of several substances totally unknown to them but apparently related to the chlorinated pesticides. At first it was thought that the compounds might be unidentified breakdown products of chlorinated pesticides. To check this idea, Dr. Jensen collected from the Swedish Museum of Natural History feathers from specimens of the white-tailed eagle for each year dating back to 1888. By this technique it was learned that the unknown compounds first occurred in feathers in 1942 and could therefore not be chlorinated pesticides or their breakdown products. It was not until two years later, after a great deal of detective work, that the contaminants were fully verified as polychlorinated biphenyls.

Part of the problem in identifying PCBs was the difficulty during analysis of separating DDT residues from those of PCBs. With analytical equipment widely used until 1965, the identity of certain PCB compounds and one of the breakdown products of DDT could easily be mistaken for each other. In Dr. Jensen's opinion it is quite likely that many DDT analyses made prior to 1965 could have been wrong. However, since 1966, with a steady improvement in analytical methods and a rapidly growing interest in the biological activity of PCBs, it has now been established that PCBs are among the most abundant, persistent, and widely dispersed environmental contaminants.

A comparison of production figures for the U.S. alone sheds some light on the relative importance of DDT and PCBs. During the period 1945 to 1968, U.S. production of DDT was about 933 million pounds. It is certain that nearly all of that production was sprayed into the environment somewhere in the world. During the years 1930 to 1970, total U.S. production of PCBs was approximately one billion pounds, of which, it has been estimated, 780 million pounds has entered the environment via the atmosphere, water systems, and dumps. Add to these figures the production of PCBs in Europe, Russia, and Japan, the total amount of PCBs entering the environment on a global basis may approach that of DDT. Japan, for instance, produced 114,660,000 pounds of PCBs between 1954 and 1971.

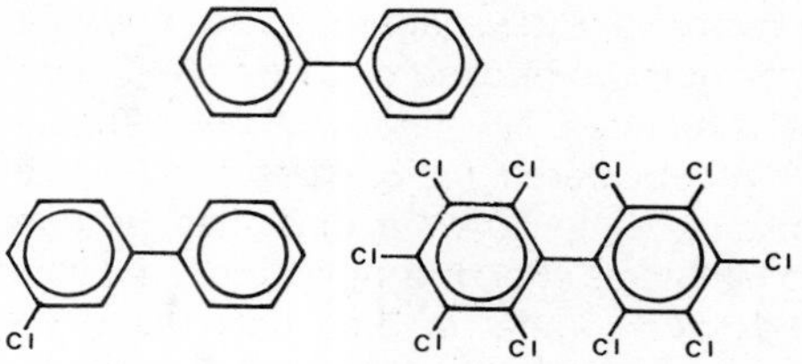

The biphenyl (sometimes called diphenyl) molecule *(top)* and two extreme examples of chlorinated biphenyls *(bottom)*. The hexagon with the circle inside represents a "benzene ring": a ring of six carbon atoms bound together, one at each corner of the hexagon. Attached to each carbon atom is either a hydrogen atom (not shown) or a chlorine atom (Cl), which makes the molecule a "chlorinated" biphenyl. A molecule with *several* chlorine atoms on the rings is called a *poly*chlorinated biphenyl, or PCB. A large number of different chlorinated biphenyls is possible, depending upon the number and positions of the chlorine atoms. Commercial PCB formulations are typically mixtures of many different compounds of this type. (Ed.)

PHYSIOLOGICAL EFFECTS

While the amounts of PCBs and DDT released to the environment may be similar, other information indicates that PCBs may be the most important in terms of ecological damage. First, PCBs seem to be far more stable—that is, because of their chemical structure they are less susceptible to being broken down by sunlight, water, and microbiological action than are the chlorinated pesticides. When DDT and PCBs are added to an activated sewage sludge and are digested anaerobically, DDT is broken down, mainly to DDD and other compounds, within a few days. PCB compounds, however, will remain intact for as long as 40 days. Anaerobic digestion in sewage sludge is one of the most rigorous techniques for determining the stability of a potential pollutant.

Another difficulty is that PCBs, rather than being one specific chemical, are a mixture of a large number of closely related but different chemicals. For example, a careful analysis of Aroclor 1254,[2] a widely used commercial PCB preparation showed that it is actually composed of 69 different chemical structures. Nineteen of these structures were considered major constituents of the mixture, while 50 were classified as minor constitutents. For the most part, analytical and toxicological studies have been conducted with commercial PCB preparations while little attention has been given the individual chemicals that make up these preparations, many of which may have unique toxicological characteristics.

Finally, while PCBs are not quite as poisonous as DDT when consumed orally, they do possess similar ability to alter important biological functions. Like DDT, PCBs have been shown to be potent inducers of liver enzymes in a number of laboratory animals. That is, when fed to laboratory animals in small doses they stimulate the liver to produce enzymes which then act on other body substrates. The clearest measure of this effect is that which occurs in PCB-treated rats when they are subsequently given a barbiturate. A measured amount of phenobarbitol, for example, will produce a standard sleeping time in laboratory rats. The sleeping time is a function of the rate at which liver enzymes break down the barbiturate to nonnarcotic substances. When rats are treated with a combination of PCBs and barbiturates, sleeping time is greatly reduced because of the increased enzyme activity brought on by the presence of PCBs in the liver.

The metabolism of barbiturates is of no importance to laboratory animals, but the enzymes induced by this effect are the same ones that metabolize certain hormones and thereby regulate important physiological functions. Their overabundance in the liver at certain times could cause serious disruptions of normal activities such as reproduction. In the past, DDT has been implicated in the disruption of the reproductive cycle of some birds, and it is thought that the mechanism has something to do with hormonal imbalance brought about by the increased production of liver enzymes at the wrong time.

PCBs have other effects on the liver. In both Japanese quail and laboratory rats, dietary levels of 100 parts per million (ppm) fed for two months resulted in enlarged livers, an increase in liver fats, and a sharply reduced liver content of vitamin A. In rats the decrease in liver vitamin A storage was 50 percent that of untreated animals.

Other subtle physiological effects have been observed in laboratory animals fed various PCB compounds. Interference with the production of adenosine triphosphatase (ATPase) in fish is a good example. ATPase is an enzyme which makes energy available for important biological reactions such as in the production of proteins, muscle contractions, the conduction of nerve impulses, and glandular secretions. At levels as low as 0.6 ppm, Aroclor 1242 inhibited the ATPase system in bluegill tissue by as much as 50 percent.

Still other effects that have been reported include edema (the accumulation of fluids) of the tissues surrounding the heart, porphyria (an abnormality of blood pigment metabolism), thickening of the skin, and a reduced resistance to certain kinds of pathogenic organisms. When ten-day-old ducklings were fed Aroclor 1254 at levels of 25, 50, and 100 ppm in their diet they suffered no apparent clinical symptoms of poisoning, but five days later, when they were exposed to duck hepatitis virus, they suffered significantly higher rates of mortality than did birds that were given the virus but were not pre-exposed to PCBs.

FOOD CONTAMINATION

While a great deal of research is still being conducted on the various sublethal effects of PCBs on laboratory animals, the accidental presence of PCBs in human food and animal feeds remains the most important aspect of their distribution. Over the past few years a number of incidents have been recorded in which significant quantities of PCBs have contaminated human food. Most of these incidents were preventable, but in some cases it is still a mystery as to what was the source of contamination.

In December 1970 the Campbell Soup Company of Camden, New Jersey, through a routine monitoring program discovered excessive PCB residues (up to 268 ppm) in chickens grown in New York State. The state of New York placed a quarantine around a three-county area and required pretesting before slaughter of all chickens originating from that area. As an interim guideline the Federal Food and Drug Administration (FDA) agreed to allow normal marketing of all chickens containing less than 5 ppm PCB. On the basis of the pretesting program 140,450 chickens were slaughtered and buried. The source of contamination in this case was thought to be plastic wrappers on stale bakery goods that were ground, wrapper and all, for use as a feed.

In July of 1971, Holly Farms of Wilkesboro, North Carolina notified personnel of the Meat and Poultry Inspection branch of the U.S. Department of Agriculture that they had been alerted to a PCB problem by a decrease in egg hatchability (a typical PCB-poisoning symptom) in their poultry operation. The source of the contamination was traced to a fish meal pasteurization plant in Wilmington, North Carolina, where a leaky heating system was dripping Aroclor 1242 into the finished product. The leak began in April 1971 and continued through July. A total of 12,000 tons of feed were contaminated, and more than 2,000 tons were recalled by the company. In the meantime, broiler and egg producers suffered tremendous losses as a result of contaminated meat and eggs. A single producer was forced to destroy 88,000 broilers, and 123,750 pounds of eggs were removed from the market.

In August 1971, Swift and Company and the Department of Agriculture notified the FDA of excessive levels of PCB found in turkeys in Minnesota. Over 100 flocks in Minnesota, North Dakota, and South Dakota were tested, but only the flock in which the original discovery was made showed high residues (20 ppm). In this case, over one million turkeys approaching market weight were held off the market until the residue levels were reduced to less than 5 ppm. The source of the contaminant was not determined.

In a routine food surveillance program, Baltimore health authorities discovered excessive PCB levels in milk, and the dairy farms involved were taken out of production by state officials. An investigation revealed that the source was spent transformer fluid (a major use of PCBs) used as a carrier for a herbicide sprayed on a power line right-of-way near Martinsburg, West Virginia.

Finally, in Ohio, Florida, and Georgia, contamination of milk was traced to a PCB-containing sealant used to seal the inside of silos. The silo coating was composed of 11 percent PCB, some of which had migrated into the silage and was subsequently consumed by dairy cattle.

YUSHO DISEASE

While North Americans have so far avoided a massive outbreak of human poisoning from PCB-contaminated food, Japan has not been so lucky. In October 1968 a number of reports of a skin disease similar in appearance to chloracne (a disease often observed in workers who handle PCBs and other chlorinated aromatic compounds) began to appear in Fukuoka prefecture in western Japan. It wasn't long before twenty other prefectures in the same general area also began reporting similar cases, and it was soon apparent that a full-blown epidemic was underway. Aside from the symptoms of chloracne, which include acne-like skin eruptions and excessive pigmentation of the skin, patients afflicted with the disease suffered from increased eye discharges, swelling of the upper eyelids, headaches, vomiting, diarrhea, fever, visual disturbances, and a number of other unpleasant effects. A team of specialists in medicine, pharmacology, agriculture, and engineering was quickly assembled from the faculties of the University of Kyushu, and the cause of the massive outbreak (1,057 patients had been reported by 1971) was soon discovered. By carefully interviewing both individuals suffering from the disease and a matched number of healthy individuals, it was found that of 60 personal traits used to compare healthy and affected individuals only one trait, that of eating fried foods or tempura daily, differed greatly between healthy and stricken people. It was also discovered that 96 percent of the diseased individuals came from households where a rice-bran oil produced by a company in Kitakyushu City and known as K rice oil was used regularly. Further investigation showed that 143 of 155 patients who had used the oil purchased it from a lot that had been produced and shipped between February 5 and 15, 1968. Finally, analysis of some of the recovered K rice oil showed that it contained from 2,000 to 3,000 ppm of Kanechlor 400, a Japanese-produced PCB that was used in the heating system at the K rice oil production plant in Kitakyushu City. Since the cause of the outbreak was discovered the disease has been known as "Yusho" or oil disease. . . .

By the summer of 1970 half of a closely studied population of 159 individuals seemed to be improving, while the other half showed no improvement, and 10 percent seemed to be worsening. There is also some evidence that babies born to women who had consumed the contaminated oil while pregnant exhibit some of the symptoms of Yusho. Thirteen women, eleven of them suffering from Yusho, and two who had no symptoms but had consumed some K oil, gave birth to ten live born and two stillborn children between February and December of 1968. Nine of the ten children showed positive signs of Yusho, and Japanese authorities plan extensive follow-up studies of the children to detect any physical or mental impairment that may develop in later years.

Ironically, six months prior to the outbreak of Yusho, Japanese health authorities had a clear warning that such an epidemic was possible. Beginning in February an epidemic of chick edema disease occurred in the same western Japan prefecture. More than two million chickens were involved and over 400,000 died. The cause of the disease proved to be PCB-contaminated feeds made by two feed manufacturers that had purchased "dark oil" for use as an ingredient in their products. The "dark oil" was produced by the same company that produced the poisonous K rice oil implicated in the outbreak of Yusho.

REGULATING PCBs

Since the repeated episodes of food contamination and the outbreak of Yusho in Japan, a number of actions have been taken both by government agencies and the major manufacturer of PCBs in the U.S., the Monsanto Chemical Company, to prevent further incidents.

Before 1971, about 40 percent of the PCB compounds produced in the U.S. were used in applications where losses to the environment were probable. These included plasticizers, lubricants, hydraulic fluids, carbonless carbon paper, and adhesives.[3] Since that time, however, Monsanto has agreed to sell PCB compounds for use only in closed systems from which spent PCBs are theoretically recoverable. The company has also constructed a high-temperature incinerator, in which Monsanto will destroy spent PCBs for customers that purchase large amounts of the materials for use in electrical transformers and capacitors. The expense of shipping the materials to the company and the cost of incineration must be paid for by the customer.

In January 1973 the National Electrical Manufacturers Association published their "official guidelines" for handling and disposing of capacitor and transformer grade PCB compounds. In the guidelines seven different locations for disposal of PCB compounds were listed, two of which have high-temperature incinerators. The other five are described as having "scientific landfills" for PCB disposal. In one facility at Sheffield, Illinois, the disposal system was described in a telephone conversation as an open pit into which contained hazardous chemicals are placed and then covered with soil. While placing spent PCBs in landfills is far cheaper than incineration at high temperatures, the security of materials handled in this way is open to question. In 1970 alone, 36 million pounds of PCBs were disposed of in dumps and landfills, and in 1972, sales of PCBs for use in transformers and capacitors were expected to be an additional 50 to 60 million pounds. Since high temperature incineration is expensive and limited (the Monsanto incinerator has an annual maximum capacity of one million pounds) it is likely that much of this production will eventually be disposed of in landfills.

To prevent further contamination of food and animal feeds, the FDA recently issued regulations placing restriction on the use of PCBs in the manufacture of food and feeds and also issued tolerance limits on the amount of PCBs that are acceptable in food as unavoidable contaminants.[4]

The new regulations forbid the use of PCBs in new equipment in food processing plants and require PCBs used in old equipment to be replaced with non-PCB-containing materials. PCBs are also to be eliminated from surface coatings and lubricants used for handling or processing food. Similar regulations were to go into effect by September 4, 1973, for plants processing or handling animal feeds.

Restrictions were also placed on equipment used to manufacture paper food containers and a tolerance limit of 10 ppm was placed on paper materials intended for use in packaging both human and animal foods. Manufacturers of plastic food containers were also included in the regulations excluding PCBs from use in equipment used to manufacture food wrappings, but no tolerance on plastic wraps was set. The regulations specifically point out that no PCBs whatever are allowable in plastic food wraps either by purposeful or accidental introduction.

The new FDA regulations, if strictly enforced, will almost certainly reduce human exposure to PCBs from accidental contamination, but they are unlikely to have an immediate effect on unavoidable contamination from environmental sources. For that reason the FDA has also established tolerance limits for foods and animal feeds in which a certain amount of PCB contamination is inevitable because of the already existing amounts of PCBs in the environment. The tolerance limits are milk, 2.5 ppm (fat), finished dairy products, 2.5 ppm (fat); poultry, 5 ppm (fat); eggs, 0.5 ppm; finished animal feeds, 0.2 ppm; animal feed components, 2.0 ppm (fish meal, and so on); fish and shell fish, 5 ppm.

One area the regulations do not cover, however, is containers and wraps for home use. Whether or not there is widespread contamination of these materials is not known, but an analysis of a number of such items in 1971 indicates that it is quite likely. The following levels were found in various plastic containers other than commercial food wraps: a plastic bag, 2.4 ppm; a plastic bottle used to contain an eyewash, 3.7 ppm; a polystyrene cup, 6.6 ppm; a baby-bottle liner, 2.7 ppm; a frozen-food bag, 4.8 ppm; a bread wrap, 3.2 ppm; a sheet plastic wrap, 3.7 ppm. None of these items are included in the regulations covering food wraps, and since none of the materials are actually plasticized with PCBs it is likely that the contamination is a result of various PCB uses in the processing machinery.

A further complication in tracking down all sources of human exposure is the fact that PCBs are now manufactured in Great Britain, France, Germany, the Soviet Union, Japan, Spain, Italy, Czechoslovakia, and the United States. It is virtually impossible at this time to determine whether or not finished consumer items that might lead to human exposure are being imported into the U.S., but the FDA plans to investigate this possibility in the future.

While both regulatory and voluntary actions were long in coming, they will quite likely reduce the immediate hazards associated with PCBs both to human beings and ecological systems. On the other hand, because of the enormous amounts of PCBs already circulating in the environment, it will be some time before they will be considered insignificant contaminants.

Editor's Note: Many of the footnotes and references in the original paper are omitted here.

REFERENCES

1. Report of a New Chemical Hazard. *New Scientist, 32:*612, 1966.
2. Sisson, D., and Wette, D.: Structural Identification of Polychlorinated Diphenyls in Commercial Mixtures by Gas-Liquid Chromatography, Nuclear Magnetic Resonance and Mass Spectrometry. *Journal of Chromatography 60:*15, 1971.
3. PCB's and the Environment. An Interdepartmental Task Force Report on PCB's. Report Number 11F-PCB-72-1, NTIS, Department of Commerce, Washington, D.C., 1972.
4. Federal Register, *38:*18096, 1973.

14

THE HYSTERIA ABOUT FOOD ADDITIVES

TOM ALEXANDER

There was a time, not long ago, when sitting down to a well-prepared meal was one of the few uncomplicated joys of life. Now, however, if a recent flood of writings can be believed, it means taking your life in your hands. The very titles of the books—*The Poisons in Your Food, The Chemical Feast, Consumer Beware, The Great American Food Hoax, Food Pollution*—are enough to knot the stomach; the writings themselves argue lengthily that we are eating our way to painful deaths and scrambling the genes of generations yet unborn. The causes of these calamities are said to be food additives, a devil's brew of unpronounceable substances cooked up in industrial laboratories.

In the last few years the case against food additives seemed to be confirmed by several highly publicized events. Quite a few food ingredients that consumers had been assured were safe were shown to be capable of causing a plague of troubles, at least in animal experiments. Monosodium glutamate, for example, could precipitate the reaction called "Chinese Restaurant Syndrome" in susceptible people—the main symptoms are burning sensations and chest pains. (The symptoms were in fact first noticed by patrons of Chinese restaurants, where large amounts of MSG are used.) Brominated vegetable oils, once widely used in soft drinks, caused heart trouble in rats that were fed high doses. NDGA, a chemical that was used to inhibit oxidation in fatty foods, produced ailments in mice ranging from cysts to kidney damage. Cyclamates caused bladder cancer in rats. And a broad array of synthetic colors, flavors, and artificial sweeteners seemed capable of causing tumors in animals. Even too much vitamin D, long used in fortifying milk, turned out to cause excessive calcium absorption in the tissues of infants. . . .

Additives, which may be defined in this context as any substances added to food in order to change its taste, texture, appearance, nutritive qualities, or "preservability," have been around a long time. Columbus

From *Fortune, 85*:62–65; 138–141 (March, 1972).
Reprinted with permission.

sailed for the Indies in search of food additives (i.e., spices). But recently there has been an enormous growth in additive use. The nationwide distribution of foods has made preservatives more important. Rising demand for premixed or precooked "convenience foods" has led manufacturers to devise chemical means of simulating the appearance, taste, and nutritive qualities of kitchen-fresh meals.

In short, the very existence and salability of many new foods is possible only because of chemicals that preserve, stabilize, leaven, thicken, emulsify, or contribute color, taste, or nutrients. Annual sales of additives to the U.S. food industry amount to roughly $500 million. The money buys more than one billion pounds of around 1,800 different substances; that's five pounds per capita.

THE LOGIC OF A CRUSADE

The industry's critics tend not to concede much value to any of these substances, even preservatives and nutrients. The preservatives are said to be there because stale foods are intended to be left on shelves too long; the nutrients because manufacturers have overcooked and overprocessed foods, thereby leaching and grinding out nature's own bounty of vitamins, minerals, and proteins. But the antichemical crusaders reserve their special ire for that large category of ingredients used solely to enhance attractiveness—including colors, flavors, stabilizers, emulsifiers, and thickeners. Taken together, the items that make foods more attractive make up roughly three-fifths of the total sales volume of additives. In the critics' eyes, they are there because the food manufacturers are out to deceive their customers about the shoddy, phony, overprocessed goods they sell. One of the critics, consumer advocate Ralph Nader, characterizes as "deception" any manufacturer's efforts to improve "palatability, tenderness, visual presentability, and convenience." By and large, most of the critics discount any possible benefits from additives or deny that these ought to be weighed in the balance with possible risks. They also tend to take it for granted that all "natural" foods are safer and more beneficial than foods containing additives. On both counts, the underlying assumptions seem arguable.

The anti-additive camp clearly has its share of cranks, conspiracy theorists, and exaggerators for effect, but there is also a deep and serious logic behind their crusade. Plenty of thoughtful scientists are concerned about the proliferation of strange new chemicals to which humans are exposed these days, and not only in food. Knowledge about the consequences has been slow to accumulate. And there have been some alarming discoveries over recent years suggesting that quite a few substances can react with, and thereby alter the composition of , the DNA molecules that make up the vital genetic blueprints for all living species. The effects of such alterations can manifest themselves in several ways:

(1) Damage to the mechanisms that limit cellular division and growth, the result being tumors and cancer. Some experts now believe that up to 80 percent of all cancer may be caused by such chemical carcinogens. Examples of chemicals that have been shown to have carcinogenic effects—in laboratory animals, at least—include the cyclamate sweeteners.

(2) Impairment of the genetic control over differentiation and development in the fetus, the result being creatures born with physical or mental defects. The chemicals capable of such impairment are called "teratogens." Some preliminary experiments suggest that the food-coloring agent Red No. 2—widely used in everything from soft drinks to baked goods—is teratogenic. The government is now reviewing the evidence.

(3) Changes in the gonadal cells that produce eggs and sperm, the result being mutated genes that are passed on to offspring, possibly creating abnormal traits. These genes can become more or less permanent constituents of the gene pool of the species. It is generally considered prudent to regard all induced mutations as detrimental; according to some estimates at least 25 percent of our health burden has genetic origins. Thus far no food additives have clearly been identified as mutagenic; however, there is a widespread assumption that most chemicals that are carcinogenic are probably mutagenic as well.

In the past, most toxicologists assumed that a single dose of most dangerous chemicals was a onetime problem: if the dose was too small to kill you, it was passed out of the system and that was the end of the problem. Now it is recognized that a good many sub-

USES AND SAFETY OF SOME FOOD ADDITIVES[1]

Additive	Function	Safety[2]	Examples of use
Adipic acid	Acidulant	Safe	Fruit drinks, gelatin desserts
Agar	Thickener	Safe	Frostings, ice cream
Algin, sodium alginate	Thickener	Unknown	Ice cream, cheese spreads
Ascorbic acid (vitamin C)	Preservative anti-oxidant	Safe	Frozen fruits, yogurt
BHA (butylated hydroxy-anisole)	Preservative, anti-oxidant	Questionable	Shortening, vegetable oil, cereal, convenience foods
BHT (butylated hydroxy-toluene)	Preservative	Questionable	Same as BHA
Calcium propionate	Preservative in baked goods	Safe	Baked goods
Calcium stearoyl-2-lactylate	Emulsifier		Baked goods, dried egg whites
Carageen (carragheenan)	Thickener	Safe for adults, questionable for babies	Milk drinks, ice cream
Carboxymethylcellulose	Thickener	Safe	Ice cream, pie filling, diet foods
Citric acid	Acidulant	Safe	All fruit drinks, gelatin
Dextrin	Thickener	Safe	Candy, powdered mixes
Dextrose	Sweetener, browning agent	Safe	Bread, soft drinks
Dimethylpolysiloxane	Antifoaming, anti-splattering agent	Safe	Vegetable oil, wine, gelatin
Disodium guanylate	Flavor enhancer	Safe	Soup mixes, canned stews
Disodium inosinate	Flavor enhancer	Safe	Same as disodium guanylate
EDTA (ethylenediamine tetraacetic acid)	Preservative, sequestrant	Safe	Salad dressing, pickles, canned vegetables
Fumaric acid	Acidulant	Safe	Pudding, gelatin, soft drinks
Glycerol (glycerin)	Moisturizer, softener	Safe	Candy, baked goods
Glycerol lactopalmitate	Emulsifier, surfactant	Safe	Cake mixes, convenience foods
Glyceryl monooleate	Emulsifier	Safe	Baked goods, pudding
Guar gum	Thickener	Unknown	Pudding, salad dressing
Gum arabic	Thickener, anticrys-tallization agent	Unknown	Cake mixes, ice cream
Hydroxylated lecithin	Emulsifier	Unknown	Baked goods, margarine
Hydroxymethylcellulose	Thickener	Safe	Ice cream, pie filling
Lactic acid	Acidulant, preservative	Safe	Frozen desserts, soft drinks
Lactostearin	Emulsifier	Safe	Cake mixes

[1]Table reprinted with permission from *Chemistry 47:8*, May 1974. Copyright by the American Chemical Society.
[2]Information from Jacobson, M. F., *Eater's Digest*, Doubleday, Garden City, N. Y., 1972.
Safe: has been subjected to full range of tests with no ill effects noted; unknown: has not been fully tested; questionable: although not proved harmful, ill effects noted in some animal studies.

stances are stored in the body tissue; thus repeated small doses can eventually build up to dangerous levels. This is the reason for much of the alarm about mercury contamination in fish, for example, as well as for the worry about pesticides and industrial chemicals like the polychlorinated biphenyls (PCB's), which, it has belatedly been discovered, are becoming ubiquitous in the natural food supply. Some scientists believe that many of the genetically active substances—the carcinogens, teratogens, and mutagens—work in the same cumulative way.

A further problem about these genetically active chemicals is that they are hard to identify. For obvious reasons, only lower animals can be used as experimental subjects; yet no other species responds precisely in the way that humans do to all chemicals. Some animals, for example, are not affected by limited doses of some carcinogens; their metabolic processes convert the chemicals to harmless substances. Furthermore, individual animals vary widely in susceptibility: it is a rare carcinogen that gives cancer to every mouse exposed to it. Hu-

Additive	Function	Safety[2]	Examples of use
Lecithin	Emulsifier, antioxidant	Safe	Margarine, chocolate, ice cream
Mannitol	Sweetener, moisture inhibitor	Safe	Chewing gum, diet food
Mono- and diglycerides	Emulsifiers	Safe	Baked goods, candy, margarine
Monosodium glutamate (MSG)	Flavor enhancer	Causes discomfort in sensitive people; general safety questioned	Soup mixes, canned stews, soups
Polysorbate 60, 65, 80	Emulsifiers	Safe	Nondairy coffee creamers, frozen desserts
Potassium bromate	Aging flour	Safe	Flour
Potassium citrate	Buffer	Safe	Imitation fruit juices
Potassium sorbate	Preservative	Safe	Cheese, jelly, mayonnaise
Propylene glycol	Moisturizer, solvent	Safe	Candy, soft drinks, marshmallows
Propylene glycol alginate	Thickener	Unknown	Frozen desserts, cheese spreads
Propyl gallate	Preservative, antioxidant	Questionable	Cereal, instant potatoes, vegetable oil
Sodium ascorbate	Preservative, antioxidant	Safe	Frozen fruits
Sodium citrate	Acidulant, antioxidant, sequestrant	Safe	Drink mixes
Sodium benzoate	Preservative	Safe, not tested for teratogenicity	Fruit juices, salad dressing, preserves
Sodium erythorbate	Preservative, antioxidant, gives red color to meat	Unknown	Bologna, frankfurters
Sodium nitrite and nitrate	Preservative, gives red color to meat	Questionable	Frankfurters and pork products
Sodium silicoaluminate	Anticaking agent	Safe	Salt, dessert topping mixes
Sodium sulfite	Prevents discoloration	Safe	Fruit juice, maraschino cherries
Sorbic acid	Preservative	Safe	Cheese, baked goods, mayonnaise
Sorbitol	Moisturizer, sweetener	Safe	Chewing gum, candy, soft drinks
Sorbitan monostearate	Emulsifier	Safe	Chocolate, frostings
Tragacanth	Thickener	Unknown	Salad dressing

mans are even more variable. Thus a chemical that has shown no harmful effects when administered to hundreds of animals of one species might still affect other species; and a chemical tolerated by millions of humans might harm others.

Confusing matters still further is the long latent period that exists between the time animals are exposed to carcinogens and the appearance of cancer symptoms. In the case of mice, this period is one to two years; in humans it can be twenty to thirty years. (This latency largely accounts for the difficulty researchers had in associating cigarette smoking with cancer.)

Finally, there is the so-called safe-dose controversy. Animal experiments show that carcinogens, like all other toxic agents, display "dose-response" relationships; that is, the higher the dose, the more animals will get cancer and the shorter the average latency. As dosages are reduced to very low levels, the carcinogenic effect likewise diminishes until at some "safe-dose" level it can be presumed that the animals will die of old age before the carcinogens give them cancer. But a safe-dose level for lower animals may be quite different from one for humans.

In addition, there are known to be synergistic effects between different carcinogens. For example, low levels of several chemicals that are safe individually can cause cancer in combination. Most tests are performed on one substance at a time. Hence anyone who

From *Sidney Harris,* American Scientist.

ventures to suggest that society might tolerate some particular levels of carcinogens does so at the risk of being charged with playing fast and loose with people's lives.

Faced with trying to legislate on the carcinogen question, Congress took refuge in the "Delaney clause" in the Food Additive Amendment of 1958. The clause, named for New York Congressman James J. Delaney, flatly prohibits the use as a food additive of any substance that "is found after tests which are appropriate . . . to induce cancer in man or animals." This is interpreted to mean that *no amount* of any known carcinogen can be added to food. And because it seems to offer a nice neat solution to some knotty questions, some people contend that the Delaney approach is the model legal principle for controlling food additives and should be extended to cover mutagens, teratogens, and many other toxic agents. Some in the harassed agencies also like the clause. "It's a regulator's dream," says Virgil O. Wodicka, director of the FDA's bureau of foods. "It relieves us of the hard decisions."

A RECIPE FOR STARVATION

While the Delaney provision seemed a model of clarity, another provision of the 1958 amendment was loosely worded. The discretion it left to the agencies charged with enforcement has been responsible for much of the current controversy and alarm. The main intent of the amendment was to ensure that the burden of proof would be on the manufacturers, i.e., before they could put an additive into food, they had to test it thoroughly. By that time, however, hundreds of additives—including such ingredients as salt, baking powder, and monosodium glutamate—were already being routinely added to food products; and very few of them had ever been tested thoroughly. To ban them until each had gone through complete testing was clearly a recipe for mass starvation.

Congress proposed to get around this difficulty by exempting from regulation substances that were "generally recognized as safe" by qualified experts. On its own, the FDA compiled lists of hundreds of the most widely used additives and sent them to hundreds of food and nutrition experts for comment. Since few of these substances had ever been tested, the comments were at best informed guesses. In any event, the famous "GRAS" list eventually came to include some 600 items. Over the years, as new techniques, doubts, and experimental evidence have come along, several items have been removed from the list. (Removal may mean either an outright ban or restrictions on the amounts of additives that may be used or on the kinds of foods they may be used in.)

Most of the deletions in the GRAS list have been made against a background of protests by the manufacturers, insisting that the additives are safe; of exaggerated warnings and accusations of bad faith from the consumerists, insisting that the additives were tolerated too long; of disagreements among scientific investigators; and of soothing noises from the government. Contemplating these scenes, the public has not been reassured by the deletions and has, in fact, become hypersuspicious of additives in general.

THE TROUBLE WITH HAM

Although not many consumer advocates will acknowledge the fact, additives have benefits as well as risks, and any reasonable policy should be based on a weighing of the two. Right now, for example, a major risk-benefit question faces the meatpacking industry and its regulatory agencies. The question refers to the chemical sodium nitrite and to a closely related substance, sodium nitrate. One or the other is employed in practically all cured-meat products, including ham, bacon, and sausages, as well as in virtually all canned meats.

These chemicals, which have been sanctioned by centuries of use, have several purposes. They have the "cosmetic" functions of reddening the meat and imparting the typical flavor of cured meats. They also prevent the development of the deadly toxins of botulism bacteria. Sodium nitrite performs both functions; even when sodium nitrate is used, it gets converted to sodium nitrite through bacteriological action, and the nitrite ends up doing the work. No substitute for nitrite is known.

Unfortunately, recent evidence suggests that, under certain circumstances the nitrite may react with a variety of other compounds called secondary or tertiary amines that are present naturally in many foods, as well as in beer, wines, tobacco smoke, and hundreds of pharmaceutical drugs. The reaction produces one of a variety of nitrosamine compounds—and some of these are formidable carcinogens. There is evidence that the nitrosamine-forming reaction can take place spontaneously in smoked fish, perhaps even in a piece of cured meat sitting on a shelf.

The meat packers are deeply worried. Cured and canned pork products amount to about 70 percent of all pork sold in the U.S.—or nearly nine billion pounds annually. Since they involve more processing than other meats, these pork products are one of the most profitable segments of the packers' business. Right now, the American Meat Institute, the FDA, and the USDA are cooperating in experiments at Swift's elaborate botulism-research labs near Chicago to determine, once and for all, whether sodium nitrite really is necessary to prevent the growth of botulism spores and, if so, what quantities are required. . . .

The question, then, is whether cured and canned meat, the staples of a $4-billion industry, must now be banned. Despite the complaints that are anticipated from bacon, ham, and sausage lovers, the answer may well be yes.

It is not surprising that the most controversial additives have been those designed to make foods taste or look better. The estimating of possible benefits from these additives is always a tenuous exercise, in which individual taste preferences invariably play a role. Critics of the food processors inevitably perceive the additives as a form of deception. The processors themselves contend that they're simply responding to consumers' wishes. And the arguments are endless. When thickening agents are used in beer, are the customers being deceived or benefited?

The processors can plausibly argue that, except for the most ascetic individuals, eating is always partly an aesthetic experience. Cooking has always been to some extent a "cosmetic" art, employing spice and other additives to change the basic flavor and texture of foods. Appearance is also important; only twenty-one years ago it was illegal under federal law for processors to color oleomargarine yellow, but they were allowed to include a dye in the package, and quite a few housewives took the time to knead the dye into the lardlike whitish mass.

Furthermore, not all the benefits of the cosmetic additives are aesthetic; their judicious use may well improve the nation's health. In 1969 a White House conference, set up to explore the paradox of widespread nutritional deficiencies in the midst of U.S. prosperity, found the answer in ignorance, traditions, and tastes. Much of the malnutrition was found in families with adequate incomes; they didn't eat what was good for them simply because they liked other things better. Nutritionists have long recognized that soybeans could provide a cheap, abundant source of vital proteins; unfortunately, plain soy products just aren't appealing to most people—unless flavor additives are put into them. Dr. D. R. Erickson, a research scientist at Swift, observes: "If given a choice between an appealing product of modest nutritional value and a highly nutritious food of only moderate appeal, people will probably purchase the more appealing product most of the time."

While the benefits of additives are often ignored, the risks are often overstated. At this point, quite a few additives have been banned because of questions about their safety; yet *there is no known case of any additive, used properly in a normal diet, having caused any illness other than the kind of allergenic reactions that many foods can cause.* Even mishaps from improperly used additives—huge overdoses taken by mistake—are so rare as to be medical curiosities.

BEWARE OF SASSAFRAS

None of this is to deny that additives may turn out to be the cause of many present ailments; obviously, we should be investigating all such possibilities. But we should also be investigating some of the natural foods that are widely assumed to be safe. The dark suspicions of additives contrast oddly with the trusting acceptance of, and failure to test, many natural foods.

The assumption that they are safe derives from the argument that mankind has been eating a variety of natural foods for millions of years, and so must have developed an ability to tolerate or detoxify the thousands of chemical compounds in such foods. But this argument is flawed on several counts. No single individual's ancestors were ever exposed to the full range of natural products that modern man uses as foodstuffs. The logic is particularly questionable in the case of the weak carcinogens—i.e., those that require repeated exposure and have long latency periods before their effects are felt. Most cancer deaths occur in old age, after the reproductive years are over; thus the process of natural selection, which might have forced humans to adapt to and detoxify weak carcinogens, has not intervened.

There is, in fact, good reason to suspect that much of present-day cancer, and some other genetic damage, is caused by foods sanctioned by long traditions. Even though research on the safety of natural foods has been limited, many have been identified as toxic. For example, an extract of sassafras called safrole—used as a flavor in root beer—was banned as carcinogenic in 1960. (Sassafras tea, ironically, is still a favorite brew of some health-food lovers.) Other natural carcinogens that have been identified include patulin (found in flour and orange juice), thiourea and several related chemicals (in cabbage and turnips), tannin (in tea and wine), and various alkaloids (in many herbs). Large quantities of nitrates have been found in many vegetables (particularly spinach) while amines are present in virtually all foodstuffs. The female hormone estrogen is carcinogenic in certain circumstances; it is found in meats, eggs, dairy products, and leafy vegetables.

Some of the most powerful carcinogens, teratogens, and mutagens are substances produced by molds and fungi. These substances include the so-called "aflatoxins," which are frequently found in peanuts, corn, and other grains under certain conditions of storage, and which may also develop in leftover foods and in bread that has not been treated with mold-retarding chemicals. When fed to a variety of animals, including rodents, poultry, and trout, the contaminated grains have been shown to cause cancers or gene mutations. (Since aflatoxins can develop in peanuts when they are still in the ground, every lot of peanuts grown in the U.S. must now undergo inspection by the Department of Agriculture.)

All of which suggests that alarm about additives may have diverted us from some more serious concerns about food. The fact is that most additives (except those on the GRAS list) have had far more testing for safety than most natural or traditionally processed foodstuffs. Dozens of familiar foods about which there is not the slightest concern would surely be banned if all foods were tested as much as additives are, and if Delaney clause standards were applied to them. With per capita consumption of additives running only five pounds or so a year, versus about three-quarters of a ton for all other foods, the additive threat by itself might be viewed as relatively minor.

WE'RE DOING SOMETHING RIGHT

It is also worth noting, in any evaluation of the threat, that the American diet as a whole is apparently becoming safer all the time. Data compiled by the American Cancer Society and the National Cancer Institute suggest that our eating habits are less likely to cause cancer than they once were. The total death rate from the disease has, to be sure, increased markedly since 1930; but most of the increase seems to be attributable to lung cancer and presumably, therefore, to cigarette smoking and air pollution. By contrast, deaths from stomach cancer have declined: from a rate of thirty per 100,000 in 1930 to about eight per 100,000 now. Some of this decline can be attributed to improvements in medical care, but stomach cancer is still fatal in about ninety cases out of 100, and so the decline almost certainly has to do with diet. And since the liver is the primary organ responsible for detoxifying substances we eat, the substantial decline in liver cancer may also be due to a healthier diet. Whether additives have had anything to do with these gains or not, we must be doing something right.

There are, in fact, several possibilities. It might simply be that a decline in the consumption of home-smoked and salt- or nitrite-preserved meat and fish products has been responsible for the gains. This hypothesis has been bolstered by data showing that Scandinavia and some other North European countries, where large amounts of smoked meat and fish are still consumed, still have high stomach-cancer rates. It might also be that better preservation techniques have diminished the amount of aflatoxins. Liver cancer is now comparatively rare in the U.S. and other Western countries, but common in parts of Africa, where diets frequently include foods containing potent aflatoxins.

Dr. John H. Weisburger, head of the carcinogen-screening section of the National Cancer Institute, is among those who worry these days about the flight from additives. "I'm not convinced that this growing trend to natural foods is a good thing," he says. "I'm not sure, for example, but that food completely free of additives may be more harmful than food properly protected by preservatives or anti-oxidants. . . ."

15
TECHNOLOGY IS FOR MANKIND

JEROME B.WIESNER

A deep mistrust of science and technology is expressed by many in our society today. Were it to prevail, this sense of suspicion and frustration could result in our failure not only to solve our present crises, some of them the result of past misuses of technology, but as well in our inability in the future to deal with problems we may not now even be in a position to predict.

The antagonism against the role of the scientist in society is very broad: it is seriously proposed that we de-emphasize basic scientific research along with technology, even that we attempt a moratorium on all new work in these fields.

Scholars such as Everett Mendelsohn, Lewis Mumford, and Herbert Marcuse claim, essentially, that modern science is a false god that must be eliminated, lest the scientific method inevitably lead to a dehumanized society, and possibly even total destruction. An increasingly large number of people, aware of the unexpected and serious side-effects of technology, express more pragmatic concerns. They suggest that most of our forward momentum in science and all of it in technology—if in fact they distinguish between the two—be halted until we have conquered pollution, urban blight, and the other frustrating problems of our day, and until we have eliminated the dangers inherent in the arms race and in our apparently rapid exhaustion of the raw material supplies of the planet.

These represent, in fact, two distinct lines of argument. The relationship of quantitative thought to humanism, a philosophic issue with which men have wrestled since even before the great Greek mathematicians, continues to elude our understanding and in fact remains unanswerable for us.

But to the more pragmatic issue of technological progress, let us be clear and unequivocal: we cannot change the way man has exploited and become dependent on his environment through his greater understanding of science and its application. To think we can make amends now by abandoning

From *Technology Review, 75:*10–13 (May, 1973).
Reprinted with permission.

scientific knowledge and technical skills is at best romantic, at least worse than futile. The fact is that many of the problems the world faces will require substantial doses of new technology if they are to be solved—sensitively relevant technology, conceived and developed with the understanding that technology can create problems, too.

On the other hand, we must also understand that even a relevant technology alone will not suffice. We need other things desperately—perhaps even more—and these involve the broadest spectrum of interests in the humanities and social sciences. We need especially to develop the ability to estimate rationally and to choose among alternate courses of action, particularly when new technology is concerned. Above all, we need the humility to admit that we will not find any absolute answers or permanent solutions.

Over the past twelve years, I have been in a unique position to feel the antagonism toward technology, and I have frequently been attacked for the many alleged wrongs of the scientific community. When I was President Kennedy's Special Assistant for Science and Technology, many critics of American science seemed to hold me personally responsible for the things they didn't like, be they the moon race, pollution, U.S. military policy, the deterioration of American cities, or the effects of television on children. (Even President Kennedy called it "your space program" when he was complaining about its cost.)

Since I became President of M.I.T., I find that I am once again held responsible for the impact of technology on our society. While neither I personally, nor M.I.T. can really take credit for creating the perplexing world in which we live, my exposed position leads me to try to understand the worries that people have and to think about what we can do to solve our problems. As the President's Science Advisor, I learned to see individual problems as parts of a large, on-going evolutionary process, and this has helped me understand what we are contending with.

MODIFYING TECHNOLOGY AND ITS EFFECTS

Two issues underlie much of the fear of technology: the widely held suspicion that most of society's serious difficulties stem from the careless or malicious exploitation of technology in the recent past; and the conviction that this exploitation will continue in the future. Moreover, since the future that is predicted is linear extrapolation of the past, the result is the doomsday prediction we hear so often today.

The fallacy of this argument, I believe, is that it ignores the considerable evidence we already have that man can in fact modify his behavior fast enough to avoid the catastrophic disasters predicted by the doomsday-sayers. To cite just a few examples:

In her book, *Silent Spring,* Rachel Carson warned of the dangers from persistent pesticides. Today, scarcely a decade later, those chemicals are severely controlled—possibly too much so—and biodegradable equivalents are on the verge of being introduced. Although many of us have already forgotten, it was also a mere ten years ago, before the partial nuclear test ban effectively halted large-scale radioactive poisoning of the atmosphere by the United States and the Soviet Union, that mothers were afraid to give milk to their children because of the strontium-90 it contained from fallout. Fifteen years ago the arms race made the danger of nuclear war very real; last year saw major steps toward nuclear arms limitation. Marine life in Lake Michigan was on the verge of extinction a decade ago; by vigorous ecological controls it has been restored.

These are but a few of the many responses that can be seen in our society. It is even possible to argue the case, as Alvin Toffler has, that much of the turmoil in the world is due to the fact that so many things are changing at once.

Obviously, it is important to listen to the critics and to try to understand them, for they are part of the process by which we learn. It is even more important to institutionalize the critical function in our society so that we need not in the future depend upon the chance appearance of a Rachel Carson or a Ralph Nader; we need to make a habit of at least trying to weigh the costs of the various choices we have before we choose one.

Most of the unexpected and serious side-effects of the application of technology—including the remarkable accomplishments of modern medicine—became major problems or threats because we failed to appreciate the power of exponential growth. In the past we responded, or did not respond, to problems as though we lived in a linear world; there seemed to be plenty of time in some distant future to correct the little troubles we preferred to ignore at the moment.

We have, I think, learned an important lesson: we can no longer charge ahead, applying technology blindly and capriciously, without coming into serious trouble. We now understand our capacity to affect our environment in all of its aspects with such power and on so large a scale that the results threaten our very existence. But even if our society *is* learning how to deal with these problems, we need to speed up this development.

This will not be easy. It will require the joint effort of people from many disciplines. It will involve new technology coupled with conscious experimentation in social process. It will involve bringing many more people into the process, and its success will require much greater general public understanding of the nature of science and of technology, of the relationships between them and their impact on social evolution. In some sense, we have been doing a good deal of this in the past: a few farsighted individuals anticipated most of the problems we face today. But society could not respond at any level—industrial, governmental or university. No one was concerned. The important fact is that we have now begun to recognize the need for coupling our foresight of technology with our effort to understand the social process.

Today a wide spectrum of citizens, industrial organizations, governmental agencies, foundations, academic institutions, and "think tanks" are aware and do care. Unfortunately at the moment our caring exceeds our understanding and our ability to manage. Much of the challenge of the next decade lies in learning how to use our technological and social capabilities and resources in a constructive and responsible manner—and to do this while still enhancing our technical capabilities.

ERROR SIGNALS FOR FEEDBACK CHANNELS

Most learning occurs through an experimental or trial-and-error process that involves selecting a goal, taking a tentative step toward it, and comparing the result with the objective. Then, if the result appears to be in the desired direction, we may take a second step. If the direction appears to be wrong, corrective action is required.

Societies can be considered large, complex, learning machines trying to satisfy the wants and desires of their citizens. What makes the situation vastly more complicated are the different goals of different members of the society and the fact that not all citizens have an equal voice in the decisions. In fact, individuals may have very different influences on different issues.

Obviously, too, the political system of a country affects in a very major way how choices are made. In free-enterprise countries such as ours, most of the decisions regarding the allocation of resources are the result of individual choices (though in every country there have always been areas such as internal and external security and education where collective actions were deemed necessary). As our society has become more complex, the number of areas where some branch of government acts for us, or inhibits our individual or group initiatives, has increased rapidly. While these trends are no doubt inevitable in an ever more complex society, they do slow down the learning process, since "feedback loops" that involve governmental action tend to be long and insensitive.

Those who have studied feedback-control systems know what happens when a system suffers from these defects. A long, slow feedback channel tends to make a system oscillate. The error signal arrives too late, with the result that the system continues to provide correction after the need for it has passed, and the controlled variable is driven too far in the new direction. Eventually, a new error signal will call for a new correction and the whole late-response effect is repeated in the other direction. (Early automobile power-steering systems suffered from this disease; we can clearly see this effect in the attempt to control the economy.) This kind of difficulty can be corrected by shortening the response time of the feedback circuit.

Or if the sensitivity of the error-detecting system is too low, only a large error is sufficient to cause any corrective action to take place. The same effect of over compensation is the likely result.

If we view our system in this light, it is

clear that we are in an interim stage, with society trying to learn how to deal with new and not fully understood problems arising from the successful application of technology. We have not yet developed adequate processes for detecting or responding to contemporary problems that require collective action.

CONTROLLING NATURE VS. CONTROLLING TECHNOLOGY

Until recent times it was not vital to do so. For most of man's history the challenge lay in coping with the natural environment, modifying or dealing with it so as to eke out a living. But within the past half century we have increased our knowledge, multiplied the forces under our control, and extended the effectiveness of our activities so much that the proper development and control of the rapidly changing synthetic environment has become as important as contending with nature.

This is what we are contending with today. Although we now recognize the need to be concerned about the total impact of large-scale exploitation of any technology and to be on the lookout for unexpected side effects which can occur from widespread and long-time uses of new processes, materials, and devices, we have just begun to develop techniques for doing this.

Our responses to some problems will probably oscillate violently for a while as we try to find the optimum way of dealing with them. For too long we were insensitive to the need to deal with the secondary problems brought on by exploiting technology; now that we recognize them, we may well overreact in many situations. Indeed, at the moment our efforts show this classical defect of a poorly designed feedback system. For example, our tardy recognition of environmental problems associated with electric power generation and our consequent inability to correct the situation rapidly has delayed construction of new power plants. As a result, many regions of the country are threatened with serious power shortages.

Those who see only the evils of technology fail to recognize that our situation would be much worse if the search for new technological solutions was stopped. The development of automotive pollution control devices illustrates this point. The technical problems are difficult and will no doubt take a long time to solve. But the alternatives would be to live with the consequences of pollution or give up automotive transportation—neither acceptable choices to most people. Meanwhile, perhaps, some intermediate steps—such as smaller cars and better mass transportation—will help.

A similar situation exists in almost all fields where the society confronts a difficulty that has its origin in labor-saving or life-expanding technologies.

The great irony of our present dilemma is that it is the consequence of success. As long as the advantages of technology benefited only a few people, they did not create large-scale environmental and social problems. To abandon these advantages to rid ourselves of the problems is hardly an adequate solution; in fact, it is obviously impossible. The great challenge is to move on from where we are to technologies and ways of employing them that will avoid uncontrollable effects in the future. Stopping science will shut off new knowledge and weaken our efforts to reverse the present situation. Technology alone is not the answer, but without technological developments few answers are likely to be found. . . .

A BROADER FRONTIER—YET ENDLESS

More than thirty years ago Vannevar Bush called science the endless frontier; it remains that today. The range of exciting research now exceeds that which was imaginable when Dr. Bush coined the title. Astrophysics and radio astronomy are showing that the vast universe in which we live is much more complex, dynamic and exciting than was suspected three decades ago. Biologists tumble over each other with discoveries about the fundamental basis of life, physicists probe the infinitesimal and increasingly understand the very nature of energy and matter, the information sciences have opened an unsuspected area of study. With these and others, mankind's growing knowledge marks a vast array of exciting research fields in which the human mind will find excitement and beauty for generations to come.

We have also discovered that the application of knowledge for mankind's benefit has turned out to be more complex and time-consuming than many people once hoped. Human societies, like matter itself, are more complex than once imagined, and every intervention has unforeseen potentials. More power has not made us wiser or more considerate. Only a better understanding of human societies combined with a growing strength from technology will help us achieve a world of decency, of increasing opportunity for individual development, and of true peace. We have only one choice: we cannot stop now—but rather must move on to a higher level of understanding, sophistication, and sensitivity in our exploitation of science, technology, and society in mankind's behalf.

PROGNOSIS

by A. Truman Schwartz

To add to the articulate expressions you have just read may be redundant. But it is the editor's prerogative, and one that I am happy to exercise. I begin by again borrowing, this time from an internationally acknowledged expert on atmospheric research, Walter Orr Roberts. In *American Scientist* for January-February 1971, Dr. Roberts wrote the following:

> We have a terrible tendency in this country to be paralyzed by imagined conflicts between technology and our traditional values. We have an awful proclivity to leave unanticipated the by-products of man's uncontrolled uses of these new tools. And when we do so, it is not the tools that are at fault but the men who use them. And this is where we must change our organizations, our social institutions, as well as our ways of doing things.

This perceptive statement seems to me to sum up the dilemmas and challenges posed by the preceding papers. By and large, the authors have been critical of technology, and justly so. Misapplied technology bears a large responsibility for our current environmental crisis. The cornucopia of progress has often produced a flood of junk—some of it merely annoying, much of it menacing. But must we settle for the solution suggested by the cartoon on the following page? I don't think so.

We must develop safe new methods of disposing of the mountains of waste described by Mr. Marx. Fortunately, technological progress is being made in this direction. Engineers are designing and testing high temperature incinerators in which industrial and urban wastes can be burned with little attendant air pollution. The energy thus evolved can be harnessed to help meet the great demands of modern society, while reducing our consumption of the earth's limited resources of fossil fuels. Moreover, such hybrid incinerator/power plants can be expected to yield valuable products while disposing of our offal. Glass, iron, copper, aluminum, and other metals can be recovered, and even the ash promises to have commercial applications. It is high time that we decided to obey the Law of Conservation of Matter and use it to our advantage.

But while improved waste disposal is a necessary step, it is hardly sufficient. We also have a collective obligation to restrict our production of affluent effluent. And here we would do well to listen carefully to Dr. Commoner and emulate Nature. Fibers that are almost indestructible may not be so miraculous after all. And the compatibility of cellulosic and proteinaceous materials with the planet's ecological cycles seems to be advantageous. Nevertheless, we would be ill-advised to abandon all research in synthetic polymers and return to the exclusive use of natural fibers. The rising cost and finite supply of cotton and wool suggest the necessity of alternative, man-made materials. Here, organic chemistry can again provide the means for creating new molecules, but this time, molecules which are more in harmony with Nature's ways.

Courtesy of *Eugene Mihaesco*

Science and technology remain society's strongest allies in combating contamination by heavy metals. Lead can finally be removed from gasoline when better methods of petroleum refining and reforming are developed, or when engines are redesigned, or when the internal combustion engine itself is replaced. Similarly, mercurial fungicides can be eliminated when safer substitutes are discovered in our laboratories. And advances in chemical engineering offer our best chances for stopping mercury loss from chlor-alkali and other plants.

A pattern is apparent in these examples: *technology is part of the solution as well as part of the problem.* To be sure, we have discovered that the bright coin minted by technology has a darkly tarnished reverse side. But to throw away the coin altogether is irresponsible stupidity. If we were to abandon the skills and knowledge of applied science in the face of impending environmental disaster, we would also abandon crops to weeds and insects, and millions of our fellow human beings to disease and starvation. This is not mere rhetoric. In the January, 1972 issue of the *Bulletin of the Atomic Scientists,* the late Eugene Rabinowitch reported that in one year, following the discontinuation of a DDT-spraying program in Ceylon, the number of deaths from malaria in that island nation rose from almost zero to one million!

As always, the solution to such a technical and moral dilemma is far easier to state than to achieve. But one thing is certain: the goal will most assuredly *not* be attained without science and technology. Replacements for PCBs, for example, cannot emerge without the creative genius of the chemist and the engineer.

Without the aid of science, we cannot even ascertain the magnitude of the problems which confront us. The very data of contaminant concentrations, often below one part per million, attest to the indispensibility of the analytical chemist and his highly sensitive instruments. And our growing knowledge of the physiological effects of minute quantities of compounds and elements exists only because of modern science. Undoubtedly, further studies of this sort will lead to a more accurate evaluation of many food additives.

It is to be hoped that society in general and scientists in particular have learned from past mistakes. We must look beyond the immediate problem and try intelligently to ascertain the long-range, widespread impact of any new product or process. When a dangerous chemical is replaced with a substitute, care must be exercised to make certain that we are not leaping from the frying pan into the fire. In some instances, acceptable alternatives may not present themselves, and for the common good we may be forced to give up a little comfort, a little convenience, and perhaps even a little freedom. And quite possibly, in the words of Theodore Fox, "We shall have to learn to refrain from doing things merely because we know how to do them."

In a democracy, to refrain from doing something or actively to pursue a specific goal is ultimately the responsibility of the entire citizenry. Such decisions seldom involve the easy alternatives of right or wrong, good or evil. Sometimes they require a choice between two goods, but more commonly they pose the dilemma of trying to select the lesser of two evils. Frequently, such hard choices must be made at the science/society interface. And after all the scientific questions are answered as thoroughly as possible, the decision is still a societal one.

Thanks to science and its applications, the prognosis is positive. If we recognize, with Dr. Wiesner, that "technology is for mankind," its power can be used to save us from suffocating in a glut of stuff and things. And then we can, with confidence, repeat these words as we look to the year 2001: "There is every reason to believe that our present slogan, 'Chemistry . . . Key to Better Living' will be justified in myriads of ways in the next quarter century."

Photo courtesy of Dr. Robert L. Wolke

SECTION THREE

MAN'S MODIFICATION OF NATURE

Drawing by Shoshana Rosenberg.

The Problem

Man's propensity for tinkering with his surroundings, with the idea of making his life more comfortable, more profitable, or just more exciting, is not new. In fact, the most successful civilizations in man's history have been those that have gained the greatest skill in shaping the environment to suit their purposes. To be sure, environmental tampering in the past has resulted in some failures, but these historical failures were small, local phenomena, not to be compared with the world-wide scale on which man can now affect the natural environment. For the first time in human history, there is now the possbility that some of the changes that man is making in Nature might be permanent and irreversible.

Under the banner of human betterment, burgeoning twentieth-century technology has produced the knowledge, the tools, and apparently even the zeal for pushing against the natural constraints of the environment until we have reached what might turn out to be the limits of its elasticity. The events of the past few years have shown us time and again that no matter how good our intentions or how seemingly straightforward our thinking, when we develop a technology to accomplish a specific purpose and when we then apply it without a clear understanding of all its possible side effects, we almost inevitably create an assortment of problems that can completely destroy its usefulness and leave us worse off than before. As we look back into the recent past by reading the following selections, we shall find that as a recurrent theme. As basic scientific discoveries are weaned from the laboratories and transformed into whiz-bang technologies, we tend to give too little thought to the question of what these applied technologies might do besides their intended purposes.

The oldest and most basic human technology of all, agriculture, is an interesting case study. Agricultural technology was born as soon as man learned that he could manipulate Nature, or at least that he could help her along in order to get better and larger yields from the soil; and as soon as he learned that he did not have to leave the sowing, cultivation and watering of his crops purely to Nature's whims. But it has been only recently (in the last half century), that what we call modern science has enabled us to produce those seemingly miraculous substances that promised at last to free man's agricultural efforts from one of his biggest problems: destructive insects and other pests which ruin his crops. These chemicals, variously referred to as insecticides or pesticides or, as R. H. Wagner labels them in Article 16, "biocides" (literally, life killers) were virtually pounced upon all over the world as the ultimate cure for one of man's most vexing problems, and they were applied to the fields in huge quantities by airplane, by tractor, by donkey and on foot. That these chemicals were, at the outset, enormously successful in accomplishing their purpose is not a debatable point. Nor is there any doubt that many of them still have a valuable place in agriculture, as Wagner points out in his article, *Biocides.* On the other hand, our tendency to rely completely on them to solve all of our pest control problems has kept us from seeing other, more balanced and rational approaches, and has entangled us in such a complex array of strategy problems—both the strategy of controlling the pests and the strategy of protecting the environment—that their overall usefulness has been greatly diminished. Thus we find that the problem of how man can best compete with pests has not been solved after all, but has merely been held at bay while the costs to both the environment and the consumer mount steadily higher and higher. In this article, Wagner portrays for us the ways in which biocides have been used and misused over the past few decades and what some of the unexpected consequences have been.

Article 17, *Biological Pest Control—Antidote to Dangerous Insecticides* by Anthony Wolff presents an alternative. Biological control is an extremely attractive alternative. It consists of fighting insect pests not with synthetic (artificial) chemicals, but with other insects or other natural enemies of the pest. The article, actually a press release from the Ford Foundation (which supports research in biological control among other things), shows that there already exists a large backlog of knowledge on methods of manipulating insect populations to achieve control with little or no use of synthetic chemicals. This concept was first demonstrated in 1899 when a group of enterprising entomologists imported a lady beetle from Australia to control an insect pest that was destroying citrus groves in Southern California. The success of this project was spectacular, and ever since that time a relatively small number of entomologists have persisted in pursuing other applications of biological control. Over the years there have been more successes, in spite of the fact that the rush to synthetic chemicals following World War II all but left biological control behind in a vast, world-wide cloud of spray.

As our society has learned, but still apparently must learn over and over again, complex problems rarely have simple, direct and all-encompassing solutions. Chemicals are certainly easy to use, but they can produce more problems than they solve; and biological control, while it can be a spectacular success, can be extremely difficult to work out and apply. Is there not, then, some sensible, balanced, middle road which can be taken to solve our pest control problems? Yes there is, but it may be even more difficult to apply. It is called "integrated control"—really more of an approach, a way of thinking, than a definite solution. In the integrated control approach, one studies all aspects of the ecological system, using biological control when possible, chemical pesticides with restraint when necessary, and even changing human agricultural habits so as to retain as much as possible the balance of Nature while shaping her operations more to our liking. Article 18, then, introduces the concept of integrated control. Titling their article *The Pesticide Syndrome,* R. L. Doutt and Ray F. Smith of the Departments of Entomology and Parasitology of the University of California at Berkeley also shine the spotlight on one more dimension of the complex pesticide problem—the ability of large industrial enterprises to turn an important ecological problem into an exercise in marketing. For profit considerations have not only determined what kinds of chemicals have been used to control pests, but in many cases have actually determined pest control *policy* over much of the world during the past 25 years. Doutt and Smith also detail some of the difficulties which arise from what they call the "closed, circular, self-perpetuating system of pest control." They show how, in severe cases, massive repeated doses of synthetic chemicals have actually led to *decreases* in agricultural yields and even to economic depression—for example, in a cotton-producing valley in Peru.

While agriculture offers the oldest and still most urgent example of man's need to modify Nature as he tries desperately to provide enough food for the earth's growing population, we can run seriously afoul of Nature's limits also in some of society's more frivolous pursuits: for example, the rush of the world's most affluent individuals to get themselves from here to there in a hurry. Of all the massively expensive technologies that have been proposed over the last three decades, few have generated as much public interest as the proposal to build a fleet of supersonic aircraft capable of flying from New York to London in three hours. While not many of us would avail ourselves of such a service, nor would many of us even need that much speed if we did, the idea was at first embraced with great enthusiasm amid vague references to national pride and the importance of maintaining the supremacy of the American aircraft industry. After all, the Soviet Union was building a supersonic transport (SST), and the British and French were building another—the Concorde, which even now flies the Atlantic—under a cooperative agreement.

Enthusiasm began to cool, however, when the enormity of the project began to sink in—both in terms of its cost versus its projected economic return and in terms of its environmental side effects. One of these side effects is detailed in Article 19. Dr. James E. McDonald, formerly with the Institute of Atmospheric Physics of the University of Arizona in Tucson, was one of the scientists who testified before the House Subcommittee on Transportation Appropriations during hearings on the SST program. The testimony at these hearings is an example of how the analytical tools of science can be applied to the critical task of evaluating possible flaws in a new technology before it gets off the ground, so to speak. Dr. McDonald was, of course, only one of the scientists who testified, and his testimony was somewhat controversial. It is important to note here that the testimony or opinion of any one scientist should not be taken as representing a consensus of the scientific community any more than should the testimony of any single witness—even a so-called "expert witness"—at a criminal trial. In scientific judgment, as with any other kind, unanimity can come only after all the pertinent facts are known. As the SST case so dramatically shows, however, far-reaching decisions sometimes have to be made long before all the experts feel that they have enough facts. While McDonald admits that his calculations showing an increase in skin cancer as a result of an operating fleet of SSTs are tentative, there appears to be a complete lack of evidence to refute that idea. The lesson here is that scientists raised the issue in a rational and scientific way, to bring to light a serious question that might otherwise never have been debated. For the time being at least, SST development in the United States has been abandoned. How much we can credit McDonald and other scientists for this outcome is not clear, but should the building of SSTs in the United States be contemplated again, the questions they have raised will have to be answered by further research and scientific evaluation, which is already going on.

In Article 20 Philip Micklin reviews what is probably one of the most serious environmental hazards that mankind has ever faced—the centuries-long task of sequestering the dangerous wastes of nuclear technology. The problem began during World War II with the development of atomic weapons and has continued to grow with the expanding civilian nuclear power program. That we have yet to find a permanent solution to this problem while we continue to push ahead with the development of more nuclear technology is a staggering realization, especially when considered in the context of the predicted growth of the nuclear power industry by the end of the twentieth century.*

We are dealing here, as Micklin points out, with materials so toxic that they must be isolated from all living things for thousands of years. So far, the best we have done is to develop a kind of nuclear juggling act—a tank storage technique that is both temporary and prone to leakage. The bizarre nature of nuclear waste materials has inevitably led to some bizarre proposals for their management. Serious consideration has been given to loading some of the more dangerous materials into launch vehicles and sending them into space. That such a seemingly wild proposal, with all its inherent risks, has been seriously discussed shows our society's complete inability to cope with this new and unfamiliar problem on the basis of any past experience.

Altogether, the following selections should give the reader some idea of man's potential for seriously and permanently altering his natural environment. As far as we can determine, we have not yet committed a major irreversible error. But the possibility of doing so is ever-present. Some of the more stable contaminants of our environment, such as the chlorinated synthetic pesticides and fission products from nuclear explosions, are now distributed over the entire globe, and we cannot say with certainty that they are of no importance, either to present or to future generations.

The kind of environmental legacy we will leave to future generations will be determined in the next few decades. What we have seen in the past three decades would be a sufficient indication that the headlong dash into technological solutions to human problems without regard for their compatibility with the total living earth is at the very least unsatisfying—and at worst extremely dangerous.

*See *The Nuclear Speedup*, page 16 (Ed.).

Kevin P. Shea

16
BIOCIDES

R. H. WAGNER

If it were possible to spray a cotton field with some miraculous compound and kill all the boll weevils without harming any other organism in the community, we would have a pesticide, a material capable of selectively killing a pest. Unfortunately, no compound is known which has this selectivity; if the boll weevils are killed, so are many other animal species, both destructive and beneficial—spiders, mites, and occasionally fish, amphibians, birds, and even mammals. So the term pesticide is somewhat general, even misleading; insecticide, while more accurate, gives no hint of potential injury beyond insects. The most general term which avoids these overlapping difficulties is biocide. If this has a grim ring to it, like genocide, perhaps it is fitting. For many people still feel that the earth would be better off without insects, mites, ticks, spiders, and all manner of "creepy-crawly" things.

Although odious in extreme usage, biocides *do* have a use and there is no intention here of insisting otherwise. We should, however, have some idea of the role that these materials play in the systems into which we introduce them; we should have a rational basis for determining at what point the cost of biocide use becomes prohibitive in economic and environmental terms. Since most of what we know about the environmental effects of biocides has come from twenty-five years of experience with DDT, many examples will deal with this biocide. But keep in mind that many environmental effects of DDT are potentially possible with *any* biologically active compound thoughtlessly introduced into the environment.

PANDORA'S PANACEA

[About] a hundred years ago, in 1874, a German Ph.D. candidate, Othmar Zeidlar, synthesized an organic compound for his dissertation, as generations of students have since done. Young Zeidlar published his work as a short note in a professional journal, took his degree, and dropped from sight.

From *Environment and Man,* Second Edition, by Richard M. Wagner. Reprinted by permission of W. W. Norton & Company, Inc.

The compound Zeidlar synthesized was not unusual, even for 1874; a series of substitutions had transformed the original ethane (C_2H_6) into a *d*ichloro-*d*iphenyl-*t*richloroethane. . . . [This compound, DDT,] sat on the shelf with a thousand other organic chemicals synthesized over the years until the late 1930s, when Paul Müller, a Swiss entomologist looking at the insect-killing properties of various compounds, found that Zeidlar's material was an extremely effective insect killer. DDT was "discovered" and for his contribution Müller was awarded a Nobel Prize in 1944.

In 1942, some DDT was brought to the United States, where the military quickly appreciated its potential usefulness. As a result of DDT, World War II was the first war where more men died of battle wounds than of typhus and other communicable diseases spread by insects. After the war, DDT was most effectively used to combat a number of insects which were carriers of thirty diseases, including malaria, yellow fever, and plague; within ten years of its first widespread use, DDT had saved at least 5 million lives. Millions of houses and tens of millions of people were sprayed with DDT in campaigns against fleas, houseflies, and mosquitoes and the diseases these carriers transmit.

Clearly a new age of chemical control of pests was underway. DDT was not the first chemical used against insects, of course. Paris green (copper acetoarsenite), first used in 1867 against the potato beetle, holds that distinction. But DDT represented the first time a totally new synthetic material was introduced into the environment on a large scale.

BIOCIDE SPECTRUM

Spurred by the success of DDT, the chemical industry explored the biocide possibilities of other biologically active chemicals.

Today there are seven main groups of biocides available to control a broad variety of plants and animals: arsenicals, botanicals, organophosphates, carbamates, organochlorines, rodenticides, and herbicides. But the three most important and dangerous from the point of view of environmental contamination are the organophosphates, organchlorines, and herbicides.

Among the first organophosphates to be marketed were parathion and malathion, which are still in use today. Their mode of action on terrestrial organisms is quite specific [and well understood. The results are] . . . spastic uncoordination, convulsions, paralysis, and death in short order. While quite toxic, organophosphates are not persistent and have much less effect on aquatic than terrestrial organisms.

Also called chlorinated hydrocarbons, organochlorines include the best known of all the synthetic poisons: endrin, heptachlor, aldrin, toxaphene, dieldrin, lindane, DDT, chlordane, and methoxychlor. Unfortunately, there is no clear understanding of just how the organochlorines work. Apparently the central nervous system is affected, for typical symptoms of acute poisoning are tremors and convulsions. Chronic levels have various effects. In aquatic organisms, which are especially sensitive to this class of compounds, the uptake of oxygen through the gills is disrupted and death is associated with suffocation rather than nervous disorder.

Another group of chlorinated hydrocarbons closely related to DDT, the polychlorinated biphenyls (PCBs), have recently emerged as an environmental threat. . . . They are [even] more stable and resistant in the environment than DDT. . . .*

Plant killers are not new. Indeed, some arsenicals have been used for years; but shortly after World War II a new group of herbicides was developed that involve plant hormones or auxins. Auxins normally are produced by the leaves of a plant and in the proper concentration serve to keep the leaves attached to the stalk or stem. In the fall, auxin levels normally drop and a layer of large, thin-walled cells forms where the leaf connects to the plant. When these cells rupture, the leaf falls. By applying a compound which lowers the auxin content of a leaf it is possible to defoliate a plant prematurely at will. This is commonly done before harvesting cotton to avoid plugging the mechanical cotton harvester with cotton leaves.

Conversely, by adding auxin at the right concentration, leaf fall and fruit drop can be inhibited, thereby decreasing loss from preharvest drop of fruit. If an excess of auxin is applied, however, plants respond by in-

*For a discussion of the PCB problem, see Article 13, page 86. (Ed.)

creasing their respiratory activities considerably beyond their ability to produce food, causing the plants literally to grow themselves to death. Strangely, the commonly used herbicides with high auxin activity, 2,4-D, and 2,4,5-T, affect only broad-leafed plants. This makes them useful tools for keeping lawns free of dandelions and other broad-leafed weeds without damaging the grass. Evidence indicates that 2,4,5-T often contains an impurity, dioxin, which can elicit birth defects similar to those caused by thalidomide; as a result, 2,4,5-T is no longer recommended for areas near human habitation in the United States.

Development of Biocides

To develop a commercial biocide, chemical companies must invest a great deal of time, research, and money. First, a large variety of materials must be screened for biological activity; toxicity is not always obvious from a structural formula. Then tests in the laboratory on a wide spectrum of pests determine toxicity levels of likely compounds. Those that are strongly toxic to one or more pests are tested for the effect of multiple doses and long-term effects, up to a month (a considerable part of the life cycle of many insects). Small plot studies by cooperating state agricultural research stations then work out problems of formulation and field dosage. Next the cooperation of farmers is enlisted for large-scale trials on a number of crop plants in different soils, climates, and regions. All of the data from these preliminary steps are then presented to the proper government regulatory agency. If approved, the compound is registered for use and marketed. Long-term toxicity tests continue as the product is used commercially. Ultimately, under this system, the consumer is the final test of any unforeseen hazard. Since the cost of marketing a biocide may run to several million dollars, none of which is returned until the product is sold, chemical companies are understandably anxious after years of development to recover their investment by promoting the use of their product. Fortunately with most large chemical companies, biocides are usually a small part of their total business, which lessens somewhat the pressure to rush a product onto the market prematurely. However in 1970 1 billion pounds of biocides were used in the United States.

Despite testing by industry and regulation by government, biocides have caused problems, some resulting from the very characteristics that made them useful pest killers. These problems can be conveniently grouped into four categories: *resistance* of the target organism; *accumulation* in food chains; *persistence* in the environment; and *nonselectivity*. Let's look at each of these problem areas in some detail.

Resistance

For a short while, DDT seemed about to push the mosquito and housefly into extinction. But by 1947 trouble developed. Italian researchers reported that the housefly was becoming resistant to DDT; corroboration soon followed from California. Soon other insects became resistant too: the malaria-carrying mosquito, then lice, fleas, and a number of other disease vectors that had formerly been controlled by DDT. Twelve insects were found to be resistant by 1948, 25 in 1954, 76 in 1957, 137 in 1960, and 165 in 1967.

Ironically, resistant pests have appeared everywhere man has sprayed, simply because the development of resistance depends upon killing as many pests as can be exposed to the selected poison. Until recently, this has been the goal of pest control.

When an insect population is exposed to a new biocide, up to 99 percent of the population is eliminated. The remaining 1 percent is by random coincidence immune or resistant to the chemical. A 99 percent wipe-out sounds almost as good as extinction. But the few resistant individuals are now able to reproduce almost without competition, a result of the elimination of both their peers and their predators. Breeding at a rate of ten generations a season, a resistant housefly population could become the most abundant strain in about five years.

Simple resistance to biocides is bad enough for the farmer, but in some instances even more serious problems were elicited. After several years' exposure to cyclodiene, cabbage maggot strains were selected that were not only resistant to the biocide, but lived twice as long and produced twice as many eggs as nonresistant forms.

A classic example of unanticipated selection for pest resistance is seen in the lygus

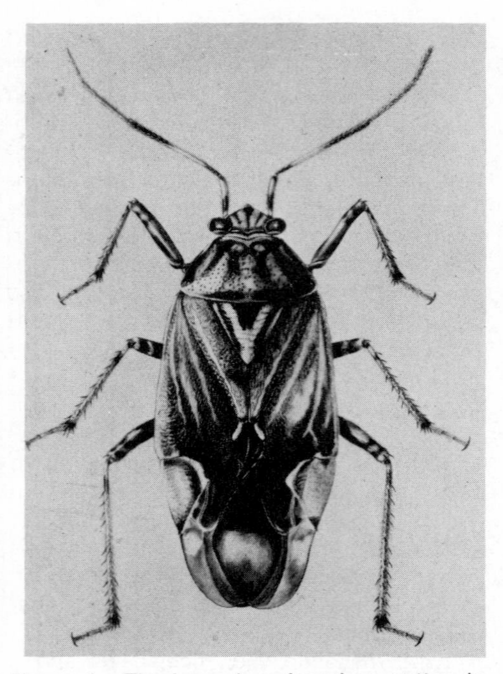

Figure 1. The lygus bug found on cotton in California may not be the pest it has been assumed to be. *(USDA photo)*

bug on California cotton (Figure 1). The lygus bug sucks plant juices from buds, blossoms, and young cotton bolls. Each one of these stages that fails to develop means one less cotton boll. In 1967, a large chemical company marketed azodrin, a new organophosphate; after much advertising, over a million acres were sprayed in California alone. While the new biocide was not persistent, it was toxic to a broad spectrum of insects and eliminated almost all insects in the sprayed fields. Its lack of persistence allowed migration of lygus bugs into the sprayed fields where their natural enemies had been eliminated (predators are always much less numerous than their prey, and therefore recover much more slowly from biocide applications). Then the lygus bug population increased and another spraying was required—then another and another. After a few seasons, it can be expected that the lygus bug will be resistant to azodrin.

The killing of beneficial insects along with the lygus bug increased the population of, and damage from, the boll worm, with the result that application of azodrin may well have decreased the yield of cotton through the increased activities of the boll worm. In addition, studies on the cotton plants showed that many buds do not form bolls and all bolls do not necessarily ripen even under the best of conditions. In fact, many buds and bolls that were attacked by lygus bugs would not have ripened anyway. In an experiment set up by University of California entomologists, the yield of cotton was not significantly different between sprayed and unsprayed plots. Hence it was quite probable that the lygus bug was not even a pest!

This type of situation can easily be repeated, because the grower, often an absentee landholder or corporation, wants someone to sample his fields occasionally, advise on the presence of potentially dangerous pests, and suggest some control. Very often the "someone" is a field salesman for a chemical company. In the cotton-growing areas of California, there are a handful of extension entomologists and over 200 salesmen representing 100 companies. Should anyone be surprised that fields are sprayed unnecessarily?

ACCUMULATION

Although DDT is only slightly soluble in water, it is quite soluble in fat. Therefore algae accumulate DDT in their cellular fat bodies [to] the [extent] of parts per million [ppm]. An organism that eats algae may accumulate . . . tens of parts per million. This accumulation and concentration of biocides in a food chain is called biological magnification. By means of this process, an extremely small quantity of a persistent biocide in water, in equilibrium with a much larger amount in the bottom mud, has unexpected effects in an ecosystem—far beyond the original intent of those who first introduced these materials into the environment. For example, bottom mud in Green Bay, Wisconsin contained 0.014 ppm DDT. But small crustaceans in the same environment had accumulated 0.41 ppm, fish 3 to 6 ppm, and herring gulls at the top of the food chain, 99 ppm—enough to interfere with their reproduction (Figure 2).

Of course, biological magnification is not limited to aquatic organisms. One early example of this process developed from the attempt to control Dutch elm disease with DDT. The Dutch elm disease is caused by a fungus that plugs the water-conducting vessels of the elm, leading to the death of the

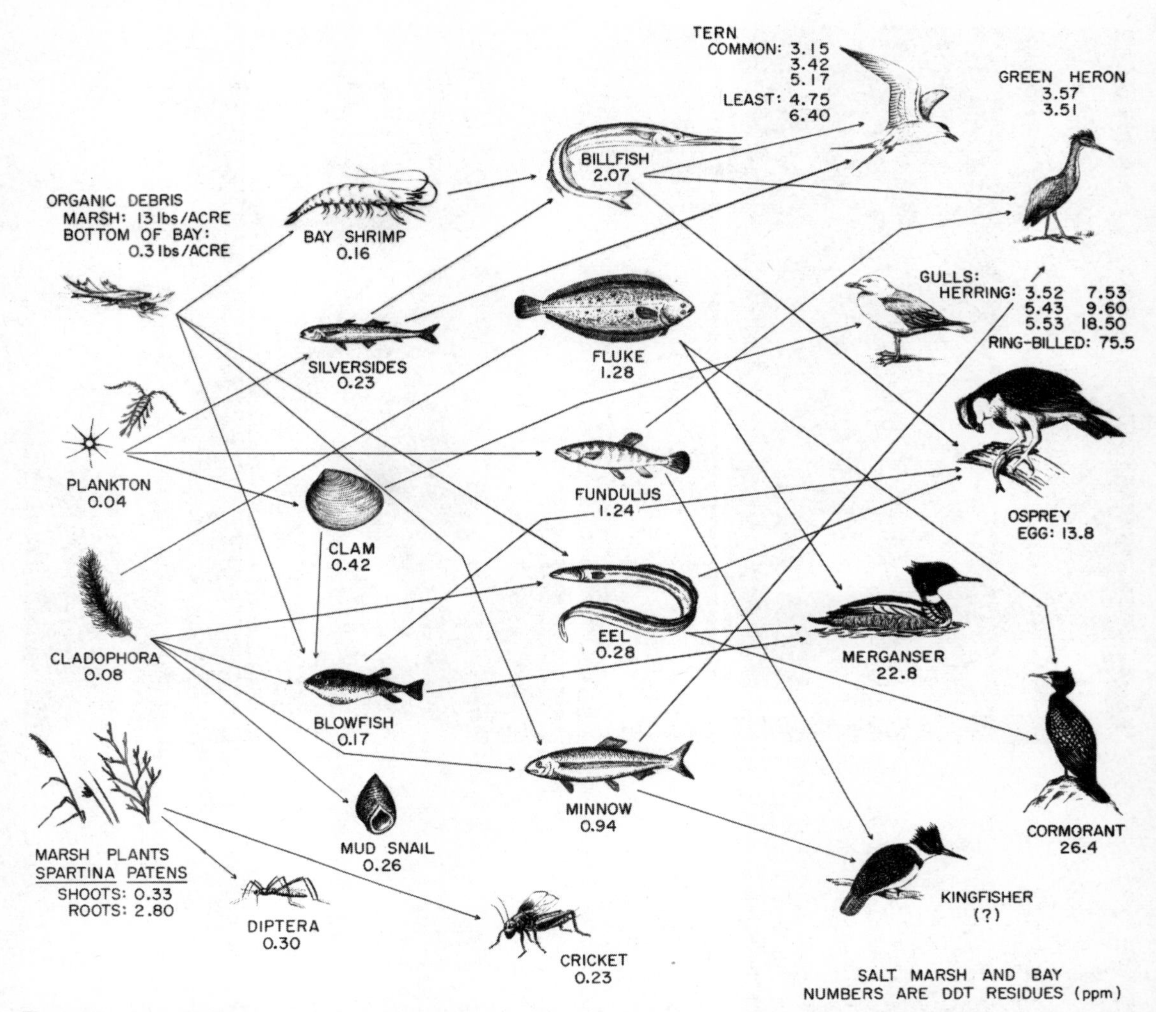

Figure 2. Food chains have been contaminated with DDT. The primary producers—phytoplankton, algae, and marsh plants—absorb DDT from mud or the water and store it in their cellular fat bodies. By passing this DDT on to the consumer, the level may gradually accumulate to large doses. *(From Brookhaven National Laboratory, Long Island, New York)*

tree (Figure 3). Since the fungus is spread by two species of bark beetles, attempts to control the disease have focused, quite naturally, on the insects. The program of spraying elms with DDT began in 1947 and until recently was still being practiced by some communities. While DDT effectively kills the bark beetles and slows the spread of the disease, it seemed in some instances to kill birds too, especially robins. The robin population on the campus of Michigan State University in East Lansing, Michigan was reported to have dropped from 370 to 4 birds in four years, and almost no nests produced young during these years. In another instance the robin population of Hanover, New Hampshire, a town which regularly sprayed its elms with DDT, fell considerably below that of a nearby town which did not spray. In both Michigan and New Hampshire, large numbers of dead birds were found to contain DDT in excess of the 30 ppm that has been observed to be lethal in robins.

In both instances, spraying seems to have been rather exuberantly carried out, applying DDT in doses much higher than necessary, thus killing more robins than usual. If DDT is applied to trees either in leaf or during dormancy, it drips to the ground beneath the trees and is accumulated by earthworms. The soil beneath a sprayed elm tree may contain 5 to 10 ppm DDT; an earthworm that

Figure 3. (a) At one time the streets of most northeastern cities and towns were lined with these airy and graceful American elms; (b) but the Dutch elm disease slowly began to kill American elms. (c) Today elms are no longer found in most towns and the stark remains haunt hedgerows and country lanes.

ingests this soil for its organic content may contain 30 to 160 ppm. A robin feeding almost exclusively on earthworms during the early spring, when elms are usually sprayed, might easily receive a lethal dose of DDT through this chain. Many apparently have. In this way a food chain can be poisoned and increasing levels of a noxious substance passed upward through the links.

The most serious and best documented example of biological magnification can be seen in the reproductive failure of certain birds of prey, especially the peregrine falcon, bald eagle, and osprey, which have undergone disastrous population decreases since 1945 (Figure 4).

The peregrine falcon has completely disappeared from the eastern United States, the

Figure 4. (a) The peregrine falcon has almost completely disappeared from the U.S. because DDT has interfered with its reproductive cycle. (b) Ironically, the national symbol, the bald eagle, is also in grave danger. (c) Even the once abundant osprey or fish hawk is rarely seen as chlorinated hydrocarbons are concentrated in the food supply of each of these species. *(Photos from Fish and Wildlife Service. a and b, photography by Luther C. Goldman; c, photography by William C. Krantz.)*

bald eagle has become rare, and the osprey uncommon. Charles Broley, who studied the bald eagle in Florida for a number of years, watched the population of eagles in one area fall from 125 nests producing about 150 birds a year in 1940 to 43 nests producing 8 young in 1957. In 1958 only 10 nests and 1 young were found. Similar decreases have been reported in osprey populations in Maine and Connecticut. In all three species eggs fail to hatch or are easily broken and eaten by the brooding bird. DDT concentrations in both the eggs and adult birds were found to be high. Falcons from an area as remote as the Northwest Territories of Canada were reported to contain 369 ppm of DDT in their fat.

How is DDT implicated in the decline of these birds? The high levels of DDT found in the fat of many birds of prey results from the same concentration process seen in the earthworm-robin chain. Birds of prey are carnivores; they eat other birds and animals which may have substantial quantities of DDT in *their* fat picked up from *their* habitats. DDT seems to affect calcium metabolism; it is, of course, the calcium deposited around the egg which produces a strong shell. This metabolic disturbance results in eggs with much thinner and weaker shells than normal, eggs that are easily broken (Figure 5). This was suggested by a study which compared the weight of bird eggs in museum collections before and after the introduction of DDT in 1947. Those birds with declining populations had substantially lighter (hence, thinner) shells after 1947 than those birds with stable populations . . . . It should be pointed out that the declining species tend to feed on species which themselves are removed by several steps from the primary consumer. The birds of prey with stable populations usually feed on the primary con-

Figure 5. DDT interferes with calcium metabolism in many birds, resulting in delicate eggshells which break easily, as in this pelican egg. *(Photo by W. Gordon Menzie)*

sumer directly. Hence they are exposed to much lower amounts of DDT. . . .

Birds tend to accumulate DDT because they cannot excrete liquids as rapidly as mammals. Some mammals also accumulate DDT, however. A study of mice and shrews in a spruce-fir forest in Maine after one pound of DDT per acre was applied to control an outbreak of spruce budworm showed that shrews had 10 to 40 times the amount of DDT residue of mice. This was due to the carnivorous diet of shrews. . . . After nine years, while the DDT residues of mice seemed to be approaching normal, the level in shrews remained unusually high. . . . Even though the level probably had little effect on the shrews themselves, what about the predators feeding on shrews? Reproductive failure has been noted in ospreys feeding on organisms with an equivalent amount of DDT residue to that found in the shrews. One can only wonder what DDT load foxes and bobcats are carrying around in forests, years after the initial introduction of DDT in their environment.

Contamination of the Ocean

It is not surprising that local ecosystems that have been sprayed with various persistent biocides pass these materials on in ever higher concentration through food chains. It was something of a shock, however, to learn that Adelie penguins on the Ross Ice Shelf of Antarctica had traces of DDT in their body fat. Although the amounts were small, 13 to 115 parts per billion (ppb) in liver tissue and 24 to 152 ppb in fat, the fact that there was *any* DDT at all, thousands of miles from the nearest possible point of contamination, was alarming. At first it was thought that the presence of several thousand scientific and support personnel nearby might be contaminating the environment in some way through discarded material, clothing, garbage, or fecal wastes, for DDT was not used for any purpose in Antarctica. But subsequent investigation in still more remote areas of Antarctica indicated that local penguins and cormorants contained 0.001 to 0.48 ppm DDE (a degradation product of DDT) and 0.011 to 0.140 ppm DDE respectively. The wide-ranging skua (a gull-like sea bird) contained 0.89 to 26.0 ppm DDE. Furthermore, . . . open ocean birds like the albatross, which touch land only to breed, were found to contain traces of various biocides. . . .

The conclusion was unmistakably clear: contamination of food chains was no longer a local affair in direct relation to biocide application at some specific point; the oceanic food chain, remote from land-based biocide usage, was contaminated. How could the enormous world ocean, stretching thousands of miles around the earth, become contaminated, whatever the scale of man's activities? Some hint might be taken from the Bravo series of H-bomb explosions in the Pacific, . . . which resulted in the catching of radioactive tuna in Japanese waters thousands of miles from the blast site. Tuna, the last link (before man) in their particular food chain, accumulated fallout to a dangerous degree. Likewise, DDT has collected in atmospheric dust hundreds and thousands of miles from dusting or spraying activities. One study showed that only 26 percent of the DDT spray intended to reach corn plants at tassel height actually made it; the rest was wafted away by wind currents. When washed out of the air by rain, the DDT can enter any ecosystem anywhere in the world. Since oceans occupy 75 percent of the earth's surface, contamination was just a matter of time. The pathway of contamination seems to be, then, land to air to ocean and perhaps finally burial in the ocean depths.

Biocides in Man

If animals can accumulate and concentrate biocides from their food, one wonders how much DDT man has accumulated from his food chain and if there is any evidence of potential harm. Man has, along with just about every other organism, accumulated DDT in his body fat. A sample of several hundred people in the United States averaged 8 to 10 ppm DDT in body fat. Whether this is or will be harmful is the subject of a sharp and continuing debate. Until 1965 there were no medically documented cases of sickness or death in man that could be traced to proper use of biocides. Of course any number of children and agricultural workers have died through careless handling of biocides, for often people do not read labels or follow instructions. Careless handling of virtually any substance, biocides not excepted, can be disastrous at times. In 1962, for example, seven infants in a hospital nursery died when salt was accidentally added to their formula instead of sugar. If, then, there is no evidence that disease or death in humans has resulted *directly* from the proper use of biocides, what are the long-term prospects?

Delayed Effects?

Apparently no one need fear inordinate residues of DDT on fruits and vegetables, for the Food and Drug Administration has set up maximum allowable levels of biocides on food products and keeps careful watch on food shipped through interstate channels. The fear of dying from one poisoned apple is as unreal as the Snow White fairy tale. But, the fact that we are all carrying a supply of DDT, regularly replenished from our environment, leads us to wonder about long-term problems caused by a body burden of DDT. There are three logical categories of DDT effects: no effects, effects that are too subtle to be correlated with DDT levels, and effects that will appear in the future. One recalls with some uneasiness the case involving a group of women who worked in a watch factory in the 1920s painting in dials with a luminescent paste containing radium. At the time, radium was regarded as harmless, so the women habitaully pointed the tips of their brushes by wetting them with their tongues. Thirty years later, many of these women began dying of mouth and tongue cancer. Today we can be appalled at the naïveté of allowing people to "eat" radium, but at the time there were no known adverse effects.

In tests where men were fed 3.5 and 35 mg of DDT a day for a period of time, no ill effects were seen. DDT accumulated in proportion to the amount consumed, then reached an equilibrium point between storage and excretion. When the dose was discontinued the body burden began to fall. Apparently the body stores DDT in response to intake rather than independent accumulation, so a person in a high DDT environment will store more DDT than a person in an environment with lower DDT levels. Although no one has died directly because of DDT properly used, many claims have been made and are being made that associate high DDT body burdens with various organ failures and malfunctions. But whether DDT directly or indirectly causes these problems or is just an innocent bystander has not been clearly demonstrated.

As long as there is any doubt about the role of DDT in human metabolism there is risk—an unnecessary risk, for we are not obliged to use DDT or any biocide. It is this doubt that led to the banning of DDT in several foreign countries and for most purposes in the United States, and will probably force restrictions on the use of other persistent chlorinated hydrocarbons as well. While there has been much obfuscation on both sides of the persistent chlorinated hydrocarbon controversy, anyone who presents an either-or choice, elms or robins, is mistaken because . . . we can have both.

PERSISTENCE

We tend to regard the soil as some kind of biological incinerator; whatever we dump is supposed to be decomposed. But some of the organic molecules like DDT are unknown in a natural environment and enzymes that can degrade them are simply not available. The half-life (the time required for half the original quantity of a substance in the environment to decompose) of DDT is around 15 years, toxaphene 11, aldrin 9, dieldrin 7, chlordane 6, heptachlor 2 to 4, and lindane 2. These are maximum values; in some environments biocides decompose much more quickly, but the figures give some idea of the time it can take for the environment to cope with these materials. Decomposition, fortunately, does not depend soely on microbial attack, or we might never get rid of some persistent organochlorines. Volatilization into the air, photodecomposition by various wavelengths of solar energy, mechanical removal in crops, and leaching are also involved in reducing levels of persistent biocides; but volatilization and microbial degradation are the most important.

The danger of persistence is best seen in the organic biocide with the longest half-life, DDT. Though only slightly soluble in water (ppb), DDT is readily absorbed by the bottom mud of aquatic systems. As DDT in the water is degraded or removed by organisms, fresh supplies from the bottom mud are released. Because of its relatively long half-life, DDT can be released continuously into an aquatic system for years.

Persistence of biocides is also found in terrestrial environments. When spray schedules are maintained over a long period of time, residue levels build up. The soil beneath some orchards in Oregon was estimated to have retained over 40 percent of the

DDT applied over a seventeen-year period. Some well-sprayed orchards accumulate 30 to 40 pounds of DDT per acre per year; others have totaled, despite degradation processes, as much as 113 pounds per acre after six years of spraying. Persistence is not limited to the newer organochlorines, though; arsenic trioxide has accumulated in some parts of the Pacific Northwest up to 1400 pounds per acre—a concentration that often poisons the crops themselves.

Some years ago the watershed around Lake George, New York was sprayed with 10,000 pounds of DDT per year for several years to control a gypsy moth outbreak. A decline in the population of lake trout followed. Although DDT residues ranged from 8 to 835 ppm in the adult trout and 3 to 355 ppm in the eggs, the adult fish were unaffected and the eggs hatched normally. Further research showed that the young fry were very sensitive to DDT, just at the time they were about to begin feeding. At DDT levels over 5 ppm there was 100 percent mortality. . . .

Perhaps the most difficult aspect of persistent biocides is the impossibility of their containment. About 15 percent of cultivated land in the United States receives biocides; 3 percent of grasslands and less than 0.3 percent of forests are treated each year. Altogether, 75 percent of the total United States land area remains untreated, and yet virtually all animals, man included, carry traces of DDT and other persistent biocides.

In response to mounting evidence from all sides indicting DDT as an environmental hazard, the Environmental Protection Agency, despite the continuing resistance of the USDA, has virtually eliminated the domestic use of DDT. The ban does not affect the export of 30 million pounds of DDT per year and a few minor, geographically limited uses may be excepted. But widespread use of DDT in the United States is apparently over.

NONSELECTIVITY

Not all of the problems caused by biocides in the environment derive from resistance, biological magnification, or persistence. Killing the wrong organism at the wrong time can be just as great a problem. One example having unexpected effects occurred a few years ago in the Near East. Jackals had been a problem for years in the settlements along the Mediterranean coast of Israel. Attracted by garbage, they stayed on to eat crops and increased greatly in number. In 1965 the government set out bait poisoned with "1080" (sodium fluoroacetate). The bait was widely distributed and no attempt was made to avoid poisoning other animals. As a result of this program, the jackals were eliminated but so were the mongoose, the wild cat, and the fox. With their predators eliminated, hares increased enormously, causing greater damage to crops than the jackals had before. Moreover, Palestine vipers, usually controlled by the mongoose, appeared in large numbers around settlements, to the consternation of the inhabitants. . . .

Intensively cultivated land is highly sensitive to the disruptive effects of nonselective biocides. Rice fields which are flooded during the growing season in Louisiana also produce red crayfish, a local delicacy which nets the rice farmer an addition to his income beyond the income from the rice crop. However, to control the rice stinkbug, toxaphene often mixed with DDT or dieldrin was used. Because of an extreme sensitivity to toxaphene, fish were sometimes killed in large numbers. Gallinules and tree ducks were also killed by eating sprayed rice. When the switch was made to malathion and parathion to avoid damage to fish and wildlife, it was found that crayfish were quite sensitive to parathion and substantial crayfish kills took place before a combination of chemicals could be worked out that endangered neither fish, birds, nor crayfish and still protected the rice.

Natural ecosystems can also be strongly affected by the application of broad-spectrum biocides, which kill insects indiscriminately. The red spider mite normally is kept under good control by a variety of organisms, especially a predatory mite. When spruce forests along the Yellowstone River in Montana were sprayed to control the Englemann spruce beetle, the spider mites, which lived in web-like structures on the undersides of the needles, were protected from the spray, but the predatory mites were killed. The following year, although the Englemann spruce beetle was controlled, there was a huge wave of spider mite damage that was worse than that of the spruce beetle. . . .

But the most ominous of all the problems generated by biocides has been the degree

Figure 6. These peach trees were being sprayed with parathion at a time (1960) when the potential ill effects of broad spectrum biocides were unrecognized. The tractor driver, without protection, was doubtless being exposed to far more parathion than the peach trees. *(USDA photo)*

to which farmers have become overdependent on them, perhaps seen at its most extreme in orchards across the country.

Spray Schedules

A hundred years ago orchards were small, containing perhaps a few dozen or at most a few hundred trees. There were enough adjoining fields and forest to allow a considerable amount of natural control. But as the orchards became larger, introduced pests could easily move from orchard to orchard, and because of the huge concentrations of trees they could inflict considerable damage. With the advent of organic sprays of low selectivity, whatever natural controls existed were reduced or eliminated, letting more pests survive and requiring more sprays. Today a typical commerical apple orchard requires a minimum of nine sprays of various biocides (Figure 6):

1. In early spring when the trees are still dormant a spray is applied to eliminate aphid eggs that have overwintered.
2. When the fruit buds begin to show green, trees are sprayed to control mildew, scab, mites, and red bug.
3. As flower buds begin to turn pink, a spray is needed every seven days to attack apple scab. There is no spraying during full flower to allow for the pollinating activities of bees, without which there would be no apples.
4. After petals fall, a spray is applied for codling moth, curculio, redbanded leaf roller, cankerworms, mites, mildew, and scab.
5. A cover spray is added a week after petal fall to get codling moth, leaf roller, cankerworm, curculio, and scale.
6. After ten more days a second cover spray is added to combat curculio, codling moth, leaf roller, fruit spot, and scab.
7. In ten to twelve days a third cover spray is added.
8. A fourth cover spray is added twelve to fourteen days later, to catch codling moth, sooty blotch, scab, and fruit spot.
9. Two weeks later, a fifth cover spray is added to kill apple maggot, leaf roller, fruit spot, sooty blotch, scab, and codling moth.

By mid-July, a second batch of codling moths is likely; then, of course, there is black rot, white rot, fire blight, and so on endlessly.

The farmer is caught squarely in the middle. If he stops spraying he is left with either no crop at all or gnarled, wormy crabapples that no one will buy. If he continues spraying, new pests will appear, old ones will become resistant, and he will sooner or later be forced out of business, unable to afford spraying twenty-four hours a day. . . .

But pest control need not always involve killing the pest. In many parts of the West, reforestation of cut or burned areas by aerial seeding is ineffective because of extensive populations of white-footed mice which devour the seeds almost at once. If mice are poisoned, new mice from surrounding areas quickly repopulate the area and eat the remainder of the tree seeds. One ingenious approach pioneered by the mammalogist Lloyd Tevis was to condition the mice to avoid all tree seeds by poisoning some with sublethal doses. Experiments indicated that one sickened mouse was worth a dozen dead ones, for once recovered, the mouse not only avoided all tree seed, poisoned or not, but taught its offspring to do likewise. Unfortunately the variations in individual mouse behavior prevented widescale use of this approach in the situation we have described, but the idea is one that might be studied further . . . .

Most of the disadvantages of biocides can be avoided. In fact, many of these materials can take their place as useful tools in the continuing war against the great variety of potential pests that man has inadvertently been nurturing through his ecosystem. [But] if mass indiscriminate spraying continues, problems seen by the late Rachel Carson will seem like bedtime stories compared with the grim realities ahead.

Several of the figures in the original article have been omitted here. (Ed.)

17
BIOLOGICAL PEST CONTROL—ANTIDOTE TO DANGEROUS INSECTICIDES

ANTHONY WOLFF

In the . . . years since *Silent Spring* made the widespread misuse of pesticides a public concern, the chemical poisons have proved as difficult to control as the pests. Although worldwide use of DDT—the original miracle insecticide—has been [more than] halved from its peak level and [has been largely] banned in the United States, it is still raining down at a rate of [many thousands of] tons per year. Moreover, since the agricultural debut of DDT after World War II, an estimated 1200 other chemical poisons in some 50,000 commercial formulations have been added to the insecticide arsenal.

The continued reliance on chemical pesticides—despite their heavy economic cost and the danger many of them pose to wildlife and human health—is due in large part to an apparent lack of alternatives. Most farmers and agricultural corporations are convinced that nothing but increasingly heavy and frequent doses of synthetic insecticides can protect their cash crops and assure food for exploding populations.

Understandably, then, the continuing pressure to limit the use of chemical pesticides has sparked a hot controversy. At a United Nations meeting in Rome [in 1971], Dr. Norman Borlaug, a Nobel Prize–winner for his contribution to the agricultural "Green Revolution," branded the worldwide movement to abandon DDT as "hysterical" and "irresponsible."

But the choice between chemicals or catastrophe offered by pesticide supporters is not the only possibility, according to a growing number of specialists in pest control. These scientists, who have worked for years in relative obscurity while the chemists dominated the agricultural stage, recommend a system of "integrated" pest control, in which chemical poisons are subordinated to so-called "biological control" techniques. "We could get by with less than half as much insecticide as we are now using," claims Dr. Robert van den Bosch, chairman of the University of California's pioneering Division of Biological Control at Berkeley.

From the Office of Reports, The Ford Foundation, 1972.

His conviction is echoed by other researchers in the field, among them Drs. C. B. Huffaker and Ray F. Smith.* Dr. Huffaker heads the International Center for Biological Control at Berkeley, which receives support from the Ford Foundation. They suggest that the use of the more environmentally polluting insecticides can be cut to one-half current levels within five years, and to one-quarter in ten, if adequate efforts are made in biological control research and education.

CONTROLLING NATURE NATURALLY

In essence, biological control preserves the integrity of natural biological systems, in which insect populations are kept in check by predators, parasites, and other environmental constraints. Biological control attempts to identify and introduce—or reintroduce—these natural factors, which are frequently eliminated from man-made agricultural environments or left behind by a migrating pest.

The potential of classical biological control was first established in California in the 1880s, when the vast groves of the citrus industry were invaded by a rampaging immigrant pest called the cottony cushion scale. The fact that the scale existed in Australia without becoming an economic pest led to the hypothesis that its population there was suppressed by some natural enemy left behind in the scale's migration to America. The identification of a predator—the vedalia lady beetle, a relative of our common "lady bug"—that feasted on the scale, and its introduction into California, led to the complete control of the pest in one year. Subsequently, the same success against the scale was achieved by similar means throughout the world.

In the decades since, according to estimates made at the University of California at Riverside, control of 223 insect pests—out of a total of 1000-or-so major candidates—has been attempted by the introduction of their natural enemies. More than half of these attempts have been at least partially successful, while complete control by purely biological methods has been achieved in seventy cases. Among the most impressive results have been the suppression of insect pests on such crops as alfalfa, olives, walnuts, grapes, and citrus in California alone. Biological control specialists claim, often with evangelic fervor, that only a lack of public faith and funds has prevented them from posting an even better score.

*See following article. (Ed.)

A lady beetle feeding on aphids. (Photo by F. E. Skinner, University of California at Berkeley.)

SNAIL-SCALE DAVID AND GOLIATH

Moreover, the biological approach has been effective in control of non-insect pests, such as the giant African snail that threatened to overruun Hawaii until a predatory snail with an immense appetite for its mischievous cousin was found. Nor is biological control effective only against animal pests. Sixty million acres of farmland in Queensland and New South Wales were reclaimed from a cactus-like weed called the prickly pear by the introduction of an Argentine moth. In another case, an imported beetle

Natural populations of cabbageworm can devastate a cabbage crop. Biological control, through the introduction of two predator parasites, enabled the damaged young cabbage plants (above) to mature into healthy No. 1 heads (below), despite the extensive injury to the older leaves. (From *Biological Control*, Plenum Publishing Corp., 1971)

was effective in rescuing some five million acres of the American West from the Klamath weed, which had made the land economically worthless.

The term *biological control* is often expanded by some advocates to include a broad and growing range of pest-control techniques that use or imitate mechanisms found in nature. The mass release of male screw worm flies, sterilized by irradiation, all but eliminated this pest in Texas and frustrated its repeated attempts at re-entry from south of the border. In other instances, a sophisticated alteration in agricultural technique—a change in planting time, for example—can reduce a pest's opportunity to cause economic damage. . . .

An "integrated" pest control program based on such biological techniques does not categorically disallow the use of chemicals. Selective use of chemical insecticides remains an important tool in the pest-control arsenal, especially in the suppression of immediate problems. However, when chemicals are indicated, an integrated program attempts to insure that they are applied with ecological economic efficiency and minimum side-effects. Thus, *integrated control* differs from present chemical control methods which spread poisons with a shotgun effect, wiping out beneficial and innocuous insects and pests alike. . . .

As evidence of the self-defeating nature of exclusive dependence upon chemicals, scientists point out that the use of a chemical insecticide often facilitates the resurgence of the very pest it is intended to eliminate by killing the pest's natural enemies as well. The chemical may also create an ecological vacuum, in which previously innocuous species can multiply and become pests. . . .

A WORLD-WIDE PROBLEM

Biological pest control . . . is claimed to offer long-term, low-cost solutions to pest problems with no negative side-effects. These advantages are becoming increasingly important as a factor in the agricultural revolution now taking place in the less-developed countries of the world. This so-called "Green Revolution" replaces native crop strains and traditional farming methods with genetically engineered high-yield varieties. While this new agriculture offers the hope of an adequate food supply for chronically undernourished millions, it also requires cheap and effective pest protection for the vulnerable exotic crop strains. . . .

. . . The Ford Foundation is supporting a special program at the International Center for Biological Control. . . . The center [is training] agricultural scientists from the less-developed countries in the theory and practice of integrated control and its application to pest problems in their native countries.

"Appropriate pest control techniques are crucial for the less developed countries," asserts Gordon Harrison of the Foundation. "The brilliant achievements of the Green Revolution clearly need to be safeguarded by redoubled attention to crop protection, which becomes increasingly difficult in greatly simplified, highly productive systems."

18
THE PESTICIDE SYNDROME

R. L. DOUTT, *and* RAY F. SMITH

The word "pesticide" is a fabricated term of recent invention. Like the vast array of chemical poisons which it appropriately designates, it is both synthetic and modern. Pesticides emerged rather slowly and in a crude form out of the nineteenth century. Then with surprising suddenness they evolved into compounds having spectacular toxicity. They proliferated enormously. They spread over the entire globe. They are characteristic of our time and have confronted us with another environmental crisis of this decade, outstanding in terms of massiveness, extensiveness and rate of change.

An engineer . . . has written that "Today we live in constant threat of man-created, irreversible phenomena. Man can literally change the face of the earth and the composition of his environment before the public and its protective agencies are aware of pending danger." This statement is particularly applicable to pesticides, for in the case of DDT the lag time between its first use and the public's general awareness of the danger it presents to the environment has been on the order of a quarter of a century. . . .

Today there is probably not a single square centimeter of the earth's surface that has not felt the impact of man. We have recently become accustomed to thinking that even though [any given square centimeter] of earth may be remote, it could have easily been caressed by our polluted air or could have been the landing pad for radioactive fallout from our diverse follies. We are not, however, quite so accustomed to realizing that this same square centimeter has very likely also received a [molecule] of pesticide. A decade ago we would have considered this entirely improbable, but now our ignorance is being rapidly dispelled for we are told that a substantial amount of pesticides enter the atmosphere either vaporized or absorbed on dust which may fall out in rain at a great distance from the area of application.

The development of these synthetic pesti-

From: "Biological Control", C. B. Huffaker, Ed. Plenum, 1971, Reprinted by permission.

Spraying DDT on an onion field to reduce loss of seed due to onion thrips. *(USDA photo.)*

cides is truly a success story. The astounding advances in modern science and technology have put these powerful chemical weapons into the bristling arsenal of all pest control practitioners. Their global use reflects the marketing abilities of corporate entities on an international scale. This is a remarkable achievement and a tribute to industrial and commercial enterprise. In the United States it is a billion dollar industry.

It is not surprising that the new pesticides had widespread and immediate acceptance. The pesticide salesman has no natural enemy in the agribusiness jungle. Uncontrolled pests can be an economic catastrophe, so the grower is constantly uneasy. His fear of loss from pests is a very real and almost a tangible thing. He is not likely to risk, through inaction, his source of income for long if he believes that his cash crop is exposed and vulnerable to attack. With very little persuasion he runs scared. In seeking protection from real or imagined dangers the grower finds that commercial pesticides are made immediately and easily available to him. An efficient sales organization is at hand to assist his purchase, even to extending credit, and in all ways encouraging the use of pesticides and furthering the myth that the only good bug is a dead bug.

Our society finds it proper that there should be reasonable profit in manufacturing, distributing, selling and applying pesticides. Nevertheless, it is unsettling in professional meetings to hear that "profit is the name of the game." It sounds vaguely unethical to an audience of research scientists, particularly those with any ecological orientation. On the other hand, the customer for many years seemed satisfied. On the occasions when a compound proved disappointing the farmer was conditioned to look for a newer, advertised as better, chemical. There was no encouragement to try an entirely different control strategy.

This led to a closed, circular, self-perpetuating system with a completely unilateral method of pest control. Alternative measures were not explored or, if known, they were ignored. The chemical cart ran away with the biological horse. Even worse was the fact that no one bothered to ask the sensible question of whether a treatment was really necessary or could be justified economically. *This we term the pesticide syndrome.* Spray schedules were based on the calendar and not on the pest populations. This incredible nonsense was masked by the prudent sounding expression "preventive treatments," and the grower was encouraged to think of this large annual expenditure as "insurance." The sales of pesticides were predictable by the calendar. This greatly simplified the logistics of supplying an agricultural district with chemical poisons and keeping these inventories in balance with the demand.

This procedure has been a very splendid, smoothly functioning commercial structure in the agribusiness community. It is unfortunate that it has suddenly become unglued. At first there were only minor hints of trouble, but these rapidly grew more alarming. Commercial chemical products, once touted as panaceas, became ineffectual on the newly resistant pest populations. This was followed by an increasingly accelerated effort by the chemical industry to develop substitutes. Cross-resistance appeared. Chemically-induced pests arose as another monstrous new phenomenon. Non-target organisms reacted in many spectacularly disturbing ways. Pest control expenses increased. This pesticide structure has continued to crumble in spite of the frantic efforts to keep it patched.

Under the pesticide syndrome none of this was at first openly admitted. Instead, it was business as usual. The gentle, though insistent, voice of Rachel Carson was met with arrogance. The stupid and simplistic question "Bugs or people?" was symptomatic of the limited thinking of those afflicted by the pesticide syndrome. Alternative methods and biological control remained as a tiny enclave in the hostile chemical world of pest control.

A change of attitude is often the hallmark of progress, and slowly attitudes began to change. The dissident murmurs from the group disdainfully scorned by industry spokesmen as "bee, bird and bunny lovers" grew into a public clamor . . . [that] has reverberated in legislative halls throughout the nation. As more and more startling facts about persistent pesticides were discovered, the formerly sacred chemical cow began to transform into an odorous old fish. This is a sad spectacle to behold, even by professionals in biological control who were so long ridiculed by the dominating chemical control proponents as a lunatic fringe of economic entomologists. It is therefore tempting to react by joining the highly vocal anti-chemical band, but biological control, even with its newly acquired aura of shabby respectability, has never lost respect for its powerful old adversary. Neither have we [who are doing research in the field of] biological control ever changed our professional opinion that pesticides when properly used are a great boon to humanity. Our philosophy, which the pesticide people never felt any necessity to adopt, has always urged entomologists to take a basic ecological approach to solving pest problems and to evaluate *all* control measures that could possibly be applied in a particular situation.

It is important to emphasize that professional practitioners of biological control today are concerned about the pendulum of public opinion swinging too far in the current anti-pesticide climate. There are indications of disturbingly negative, restrictive and extremist positions being taken against pesticides. There are anti-pesticide crusaders and hordes of "instant ecologists" who have appeared to bury the chemical Caesar, not to honor him. There are some strange political undercurrents and an uncommon amount of emotion involved. All of this simply increases the difficulty of finding sensible solutions to the pesticide dilemma. . . .

Disregarding the sound and fury that has been generated over the pesticide issue, the fact remains that problems with their use in agriculture, public health, and forestry are of such a magnitude that almost everywhere a reappraisal is necessary. Actually this has quietly been under way for a number of years in many parts of the world, so there is now a substantial amount of experience with alternative paths and from this a very clearly defined concept for future research programs has arisen. . . .

The strategy of pest population management, as opposed to that of preventive treatments, offers the most promising solution to many of the present difficulties. This strategy stems from the notion that pest control is adequate if populations are held at tolerable levels. The unrealistic idea that pest control must be equated with total elimination is rejected. It is recognized that the most power-

Spraying a lettuce field by plane. *(Shell photograph, courtesy of U. S. Department of Agriculture.)*

ful control forces are natural ones, some of which can be very effectively manipulated to increase the mortality of a pest species. The pesticide is considered as the ultimate weapon to be held in reserve until absolute necessity dictates its use. This is a sophisticated use of ecological principles, and is both intellectually satisfying on theoretical grounds and extremely effective in application. This in general terms is the concept known and practiced as *integrated control*.*†

*Editor's italics.

†A Food and Agricultural Organization of the United Nations (FAO) panel of experts defined integrated control as a pest management system that, in the context of the associated environment and the population dynamics of the pest species, utilizes all suitable techniques and methods in as compatible a manner as possible and maintains the pest populations at levels below those causing economic injury. It emphasizes the fullest practical utilization of . . . existing . . . factors in the environment; it is not dependent upon any specific control factor but rather coordinates the several applied techniques appropriate for the particular situation, with the natural regulating and limiting elements of the environment.

There is still considerable misunderstanding of what is meant by integrated control. It is not merely a workable marriage of chemical and biological controls. Its use goes far beyond this. The best features of all control measures are utilized and applied within the framework of practical farm operations. There are several unique aspects to the research upon which an integrated control program is built. One is philosophical since the researchers must believe that chemical treatments ought not to be applied until they are clearly needed. Another aspect is organizational in that the program is best developed by a team of specialists representing diverse disciplines but all of whom are channeling their research toward a common goal. The third and fundamental aspect is that the program is based on a sophisticated understanding of the ecology of the ecosystem involved. . . .

All ecosystems are extremely complex

and [are] composed of many sub-systems . . . . It is usually extremely difficult to separate one component and evaluate its role and its influence in the system but this can be done, and now is being done in integrated control programs. . . .

[An example] of the development of integrated control programs and the problems involved [is] found in the . . . control of cotton pests in Peru. . . .

For cotton, the pattern of crop protection may go through a series of five phases, or it may stop at a particular phase, or cotton production may cease before all phases are established. The sequence can be described as follows:

1) *Subsistence Phase.* The cotton, usually grown under non-irrigated conditions, is part of a subsistence agriculture. Normally the cotton does not enter the world market and is usually processed by native hand weavers. . . . There is no organized program of crop protection. Whatever crop protection is available results from natural control, inherent plant resistance in the cotton, hand picking, cultural practices, rare pest treatments, and luck. . . .

2) *Exploitation Phase.* In many newly irrigated areas in developing countries, cotton is often one of the crops first planted to exploit the new land resource. Crop protection measures are introduced to protect these large and more valuable crops. These measures are justified to take full advantage of the irrigation scheme and to maximize yields. Crop protection methods are sometimes needed to protect a new variety having qualities of increased yield and better fiber but poorer insect resistance. Unfortunately, in most cases, in the exploitation phase the crop protection measures are dependent solely on chemical pesticides. They are used intensively and often on fixed schedules. At first these schemes are successful and the desired high yields of seed and lint are obtained.

3) *Crisis Phase.* After a variable number of years in the exploitation phase and heavy use of pesticides, a series of events occurs. More frequent applications of pesticides are needed to get effective control. The treatments are started earlier in the cotton growing season and extended later into the harvest period. It is notable that the pest populations now resurge rapidly after treatment to new higher levels. The pest populations gradually become so tolerant of a particular pesticide that it becomes useless. Another pesticide is substituted and the pest populations become tolerant to it too, but this happens more rapidly than with the first chemical. At the same time, pests that never caused damage in the area previously, or only occasionally, become serious and regular ravagers of the cotton fields. This combination of pesticide resistance, pest resurgence, and the unleashing of secondary pests, or the induced rise to pest status of species previously innocuous, occasions a greatly increased application of pesticides and causes greatly increased production costs.

4) *Disaster Phase.* The heavy pesticide usage increases production costs to the point where cotton can no longer be grown profitably. At first only marginal land and marginal farmers are removed from production. Eventually, cotton is no longer profitable to produce in any parts of the area. A number of Central American countries are now in this disaster phase. In the United States, this disaster phase has been postponed a bit by cotton subsidies.

5) *Integrated Control or Recovery Phase.* In only a few places has the integrated control phase been developed. The most noteworthy examples are in several valleys in Peru. In the integrated control phase, a crop protection system is employed comprising a variety of control procedures rather than pesticides alone. Attempts are made to modify the environmental factors that permit the insects to achieve pest status, and the fullest use is made of biological control and other natural mortality.

The pattern outlined above can be seen in the case history of cotton pests in the Canete Valley of Peru. . . . In the 1920's the Canete Valley shifted from an emphasis on cultivation of sugar cane to cotton. Approximately two-thirds of the valley is devoted to cotton. . . . The Canete story is a classic example of the pest control problems that can beset the grower if he ignores ecology and relies unilaterally on chemical pesticides.

During the period from 1943 to 1948, the chemical control of cotton pests was based mainly on arsenicals and nicotine sulphate, although some organic chlorides were used toward the end of the period. The main insect pests, a "gusano de la hoja" (*Anomis texana* Riley), the "picudo peruano" (*Anthonomus vestitus* Bohm), and the "perforador pequeño de la bellota" (*Heliothis virescens* F.),

"There I was, coming in low at a hundred and seventy-five m.p.h. with forty-two acres of broccoli to the left of me, eighteen acres of asparagus to the right of me, eighty acres of carrots straight ahead! Power lines all around! My target: seven acres of badly infested garlic smack in the center. . . ." (*Drawing by Dedin;* © *1972,* The New Yorker Magazine, Inc.)

gradually increased until there was a severe outbreak in 1949. It was presumed that this outbreak, and the associated heavy aphid infestations, were brought on by the widespread use of ratoon cotton (second and third year cotton) and the introduction of organic insecticides. . . .

The next period lasted roughly from 1949 to 1956. Ecologically oriented entomologists . . . recommended a variety of cultural controls and a return to inorganic and botanical insecticides. This advice was not followed in full. One-year ratoon cotton was permitted and a heavy reliance was placed on the new exciting organic insecticides (mainly DDT, BHC and Toxaphene). Some cultural practices were modified to increase yields. New strains of *Tanguis* cotton were introduced and more efficient irrigation practices initiated.

These procedures were very successful initially. Yields nearly doubled. . . . Farmers were enthusiastic and developed the idea that there was a direct relationship between the amount of pesticides used and the yield, i.e., the more pesticides the better. The insecticides were applied like a blanket over the entire Valley. Trees were cut down so the airplanes could easily treat the fields. Birds that nested in these trees disappeared. Beneficial insect parasites and predators disappeared. As the years went by, the number of treatments was increased; also, each year the treatments were started earlier because of the earlier attacks of the pests. In late 1952, BHC was no longer effective against aphids. In the summer of 1954, toxaphene failed to control the leafworm, *Anomis*. In the 1955–1956 season, *Anthonomus* reached high levels early in the growing season; then *Argyrotaenia sphaleropa* Meyrich appeared as a new pest. *Heliothis virescens* then developed a very heavy infestation and showed a high degree of resistance to DDT. Organophosphorus compounds were substituted for the chlorinated hydrocarbons. The interval between treatments was progressively shortened from a range of 8–15 days down to 3 days. Meanwhile a whole complex of previously innocuous insects rose to serious pest status. . . . In nearby similar valleys, where the new organic insecticides were not used, or only in small amounts, these insects did not become pests. Finally, with most of the pests resistant to the available pesticides and the insects rampant in the cotton fields, the 1955–1956 season was an economic disaster for the growers of the Canete Valley. Millions of bales were lost. In spite of the large amounts of insecticides used in attempts to control the pests, yields dropped. . . .

The development of integrated control followed this severe crisis. When the Canete growers realized that they could not solve their pest problems with insecticides alone, but had, instead, produced a catastrophe, they appealed to their experiment station for help. A number of changes in pest control practices were made, including certain cultural practices. Cotton production was forbidden on marginal land. Ratoon cotton was prohibited. The Canete Valley was repopulated with natural enemies from neighboring valleys and beneficial forms were fostered in other ways. Uniform planting dates, timing and methods of plowing, and a cotton-free fallow period were established. Use of synthetic organic insecticides in specific fields was prohibited except by approval of a special commission. There was a return to arsenicals and nicotine sulphate. These changes were, with the approval of the growers, codified as regulations and enforced by the Ministry of Agriculture. As a result of this new integrated control program, there was a rapid and striking reduction in the severity of the cotton pest problem. The whole complex of formerly innocuous species, which cropped up in the organic insecticide period, reverted to their innocuous status. Furthermore, the intensity of the key pest problems also diminished and so there was an over-all reduction in direct pest control costs. By the next year, the yield was back to [its 1943–1948 level] and since has [increased to] the highest yields in the history of the Valley. . . .

CONCLUSION

This experience in Peru [and a similar one with grapes in California] strikingly illustrate that pest control cannot be analyzed or developed as an exercise in control of a given pest alone; rather, it must be considered and applied in the context of the ecosystem in which a number of pest, or potential pest, populations exist and in which control actions are taken. Whatever the new technologies which may be developed for control or management of pest insects in the future, they must be applied with a knowledge of the operation of the pertinent ecosystem. This means that in addition to the fundamental studies on the new methodologies, a series of parallel investigations should be carried out to provide a solid base of information on the significant factors in the ecosystem influencing the pest populations and possible potential pests. Both of these case histories show how quickly an agricultural industry can recover from the pesticide syndrome, and they offer hope to others afflicted by this malady so peculiar to the 20th century.*

*References in the original article have been deleted. (Ed.)

19 ENVIRONMENTAL EFFECTS OF THE SST: STATEMENT TO THE HOUSE SUBCOMMITTEE ON TRANSPORTATION APPROPRIATIONS*

JAMES E. McDONALD

INTRODUCTION

Deliberations on the national decision as to whether to proceed with an SST program have brought into public debate important questions concerning possible environmental effects of major SST fleet operations. I wish to summarize here some points which I think have been overlooked or underemphasized in recent discussions of the pros and cons of initiating a major air-transport technology in the stratosphere.

During the past several months, a substantial share of my time has been spent in assessing certain specific questions as to how SST fleet operations might affect the earth's atmosphere and thereby modify either climate, weather, or human activities contingent upon important atmospheric processes. Although my work was undertaken in connection with current studies of the National Academy of Sciences' Panel on Weather and Climate Modification, I wish to make clear that I do not speak here for our Panel, but rather as an individual scientist. However, I wish at the same time to express my indebtedness to many other scientific colleagues in various parts of the country who have offered advice and critique in various phases of these analyses. Although some of my viewpoints and findings are still tentative, all have been laid before a substantial total number of workers in a variety of different fields in an effort to detect and eliminate gross errors; so I acknowledge the help and criticism reflected in what I shall say here.

Before examining in some detail one specific SST atmospheric effect and its consequences (ozone reduction and its effect on skin cancer incidence), I want to emphasize three principal generalizations that strike me as having strong bearing not only on the present SST decision but also on the feasibility of moving on even further toward a still higher-altitude mode of air transport now under engineering study, namely the HST (hypersonic transport) technology.

*Hearings of 2 March 1971. These hearings, among other considerations, led to the suspension of the United States' plans to construct supersonic transport (SST) airplanes for commercial passenger service. The French-British SST, the Concorde, has, however, been built and flown. Dr. McDonald's testimony, reprinted here with minor changes of wording and with references deleted, was not universally subscribed to by other scientists; certain of the conclusions expressed therein have since been disproved and others are still not completely resolved. It is presented here only as an exceptionally clear example of the way in which expert scientific opinion can be elicited and weighed before a society embarks upon a new technology having complex and possibly serious environmental implications. (Ed.)

To make clear certain points that will come up repeatedly below, it needs to be noted first that the present generation of *subsonic* jets, such as the 707, DC-8, and 747, which cruise at altitudes near 35,000 feet, are still operating within the *troposphere,* the lowest major subdivision of the atmosphere, separated by the thin *tropopause* layer from the next higher and quite different region, the *stratosphere.* For present purposes, we may take the mean altitude of the tropopause as about 40,000–45,000 feet over middle latitudes. SSTs and HSTs would, by contrast, operate well within the stratosphere. The proposed Boeing SST would cruise near 65,000 feet, while the HSTs, still only on drawing boards, would cruise at altitudes that might begin near 80,000 feet, working upwards as technology advanced to ultimate HST cruise levels perhaps near 150,000 feet.

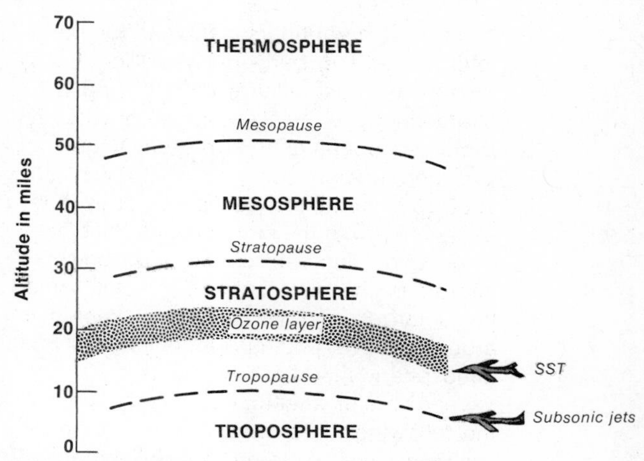

The earth's atmosphere, showing the approximate limits of the various regions.

Although there are military jets (such as our U-2 or our SR-71) that can cruise at stratospheric altitudes, and although a number of interceptors have short-duration altitude capabilities of well over 50,000 feet, the total

One of the designs which had been proposed for the U. S. supersonic transport (SST) program, now cancelled. *(Courtesy of the Boeing Commercial Airplane Company.)*

number of flight-hours per year (or better, total tons of fuel burnt per year) logged by present and past military aircraft flying in the stratosphere is small, compared with the projected operational levels envisaged for U.S. and foreign commerical SSTs by the 1980–85 period. Hence we do not yet have experience with the environmental effects of flying large numbers of very high powered aircraft in the stratosphere, a point sometimes forgotten or even misrepresented in arguments over potential seriousness of SST atmospheric modification effects.

Very careful attention should be paid to the following three points, as we weigh the pros and cons of the present SST proposal:

(1) The stratosphere is effectively about 100 times more sensitive to technologic contamination than is the troposphere because its turnover time averages about 100 times longer than that of the troposphere.

(2) The stratosphere, unlike the troposphere, in which our current air transport technology now operates, is a region of high chemical reactivity.

(3) If we now start an SST transport technology and then later attempt to improve range efficiencies by modifying engine or airframe designs to permit flying at still higher altitudes (or if we move on to an advanced HST technology), then we shall find that both of the preceding difficulties grow even more serious the higher we try to fly in the stratosphere.

The *first* of those three generalizations hinges on the important quantity known as the *mean turnover time* (also called the residence time, exchange time, or holdup time). In the troposphere, into which we emit essentially all of our present industrial and technologic pollution, the effective turnover time for the major pollutants averages only about a week, and possibly less than that. Precipitation processes rapidly scavenge particulates and many gaseous pollutants from the troposphere, so that contaminants (even those from present-day jets flying in the upper troposphere) are fairly quickly washed out by rain. In the stratosphere there is no efficient scavenging action; it has no cloud-and-rain washout mechanisms comparable to those that are effective in our troposphere. Instead, removal of gases or particulates emitted into the stratosphere hinges upon slow transport and downward mixing to the tropopause, followed by "tropopause folding" or other leakout mechanisms that carry the pollutant down into the troposphere where rain scavenging can complete the removal process. For the lower stratosphere, where the proposed SSTs would fly, the turnover time is now regarded as averaging about two years, in contrast to the troposphere's five to six days. Hence, for any given pollution rate, residence times run about 100.times greater in the stratosphere because contaminants take 100 times longer to be flushed out. Actually that ratio of 100-fold should be set at an even higher figure, since the great stability of the stratosphere prevents the kind of deep mixing characteristic of the far more unstable troposphere, with the result that the mass-thickness of stratospheric air effectively available to dilute contaminants is about five times smaller than for the troposphere. Without pursuing these matters into further detail, we may say that we are probably *underestimating* the seriousness of this point by here adopting the figure of 100-fold greater turnover time in the stratosphere where the SSTs would be emitting various exhaust products.

The *second* point, concerning the far greater degree of chemical reactivity of the stratopshere, results from the presence there of an *ozone layer,* from the presence of small but chemically quite significant concentrations of reactive '*free radicals*" such as hydroxyls and peroxyls, and from the presence of an intense flux of solar ultraviolet radiation, whose energy is sufficiently high to drive many chemical reactions that cannot occur in our lower atmosphere. (The ozone layer filters out much of the solar energy before it reaches the troposphere.)

The *third* generalization calls attention to the fact that if future aeronautical improvements in the SST should permit flying at higher altitudes (which is attractive for reasons of both fuel economics and sonic boom mitigation), it would tend to exacerbate the foregoing two difficulties. Average turnover times increase at higher altitudes in the stratosphere (a trend well-documented from nuclear bomb-test debris tracer studies), probably attaining values of the order of 10 years near 100,000 ft. This implies a tendency to build up still higher steady-state concentrations for any fixed rate of injection of aircraft contaminants as flying altitudes are raised from the currently proposed 65,000-ft cruise level. This difficulty is further aggravated by the fact that the main peak of the ozone layer lies above the 65,000-ft level. (Ozone concentrations attain a broad maximum centered in about the 75,000–100,000-ft interval, the exact value depending somewhat on latitude and season.) Relative concentrations of free radicals and intensities of reaction-energizing solar ultraviolet radiation increase as altitude increases, and would reach even higher levels at altitudes such as those now being considered for HST operations. If there are (as I believe to be the case) some serious environmental consequences of starting an SST technology at 65,000-ft cruise levels, those difficulties will get worse as efforts are made to push SST cruise levels still higher. And, without going into the point in full detail, there will certainly be pressures to push for those higher altitudes, since it is believed that fuel consumption will be steadily lowered as cruise altitudes and speeds are increased. Longer holdup times at these higher levels will tend towards higher and higher steady-state contaminant concentrations and, at the same time, the greater amount of solar ultraviolet radiation will tend to pose steadily more serious problems of environmental side effects. This point is not, I believe, widely appreciated in the aeronautical engineering world.

These three rather broad generalizations, I submit, need to be weighed very carefully in any major national decision to undertake an SST technology. They have been ignored in a number of recent defenses of the SST program, with the result that SST exhaust emissions are sometimes viewed as being of minor concern. The full implications of the long holdup time and high reactivity of the part of the atmosphere in which SSTs would operate have only begun to be explored.

Furthermore, there is a *fourth* generalization one might well append to the above three: Increases in the concentrations of stratospheric particulate concentrations resulting from SST exhaust emissions and their reaction products (sulfates, nitrates, soots, hydrocarbon products) will be introduced into a part of the atmosphere where they may be able to exert more climatically adverse effects than similar particulates would in the lower atmosphere. This is a subtle point (and one entailing some still poorly known optical properties of possible particulates resulting from SST stratospheric processes), so I shall mention it only in passing—but, at the same time, it is necessary to warn that long-term operation of new types of high-altitude air transport technologies might make this still poorly understood form of environmental disturbance as serious as, or even more serious than, any others now suspected.

SOME FALLACIES AND MISUNDERSTANDINGS ABOUT POTENTIAL SST ATMOSPHERIC EFFECTS

(1) *SSTs will cause persistent ice-crystal veils which will alter the earth's climate.* Fears that the extremely large volumes of water vapor emitted from SST fuel combustion will lead to contrail formation and hence to development of long-persisting hazy stratospheric veils of slow-falling ice crystals do not appear to be well founded. I have reviewed again the arguments that led me and others who prepared the 1966 National Academy of Sciences report on weather and climate modification to discount this problem and have found no reason to alter our 1966 conclusions. SSTs will not only fly at altitudes too high even to lead to the formation of contrails most of the time but, still more to the point, they will be flying in a region where mean relative humidities due to naturally occurring water vapor average only about five percent. Persistence of any contrails that do occasionally form is ruled out by such dryness. Only rarely, at high altitudes in the winter, or possibly occasionally at low altitudes, is it at all probable that contrails would form and persist. Serious climatic disturbance from these comparatively rare occasions does not appear likely.

(2) *Water vapor additions to the stratosphere will produce such tiny reductions of ozone that no biologically serious consequences will ensue.* This conclusion was reached in the SCEP Report*, and it was also my own initial conclusion. However, a previously overlooked line of evidence now appears to lead to quite opposite conclusions. The possibility that ozone reduction resulting from chemical interactions with SST-exhaust water vapor will be large enough to yield serious increases in incidence of skin cancer over the estimated

---

**Study of Critical Environmental Problems; Man's impact on the Global Environment,* Cambridge, MIT Press, 1970.

operating period of an SST technology is a good example of a subtle and initially unrecognized environmental hazard now calling for the most searching scrutiny. I shall take up this example of a "hidden SST problem" below and use it to show how there may well be difficulties in high-altitude transport technologies that we have only barely begun to understand. Briefly, it is my present estimate that operation of SSTs at the now-estimated fleet levels predicted for the period from 1980 to 1985 could so increase transmission of solar ultraviolet radiation as to cause something on the order of 5000 to 10,000 additional skin cancer cases per year in just the United States alone. (I return to this point below.)

(3) *Water vapor added to the stratosphere by SSTs is of only trivial significance, since thunderstorms put far more water into the stratosphere by entirely natural processes.* This argument has been heard often. In one widely repeated form, it suggests that a single tropical thunderstorm can inject as much water vapor into the stratosphere as would the entire SST fleet in a single day. The argument then usually continues with the remark that, for the world as a whole, there may be something like 4000 thunderstorms per day, hence why worry about SST water vapor additions? The primary fallacy here is that any and all *natural* processes (including thunderstorms) accounting for the naturally occurring water vapor in the stratosphere are already fully allowed for in the present estimate of the *average* natural water vapor content of the stratosphere (about 5 ppm by volume). I know of no analyses of potential SST environmental effects related to SST exhaust vapor contamination of the stratosphere that have not proceeded from just this basis; hence I can only regard the "thunderstorm argument" as inherently misleading because it makes the uninitiated think that SST vapor additions are somehow trivial when measured against wholly natural effects. The important consideration is that SST flights *will* result in an additional water vapor burden in the stratosphere.

Furthermore, the SST proponents who use the thunderstorm argument have, to my knowledge, never backed up their basic claims with good observational data as to what fraction of the several thousand thunderstorms per day actually penetrate the tropopause and succeed in delivering any vapor to the dry stratosphere, and as to just how much vapor is actually exchanged with the stratosphere even in those cases where a thunderstorm does penetrate the lower stratosphere. I would suggest that only an extremely small percentage of all thunderheads (*i.e.,* cumulonimbus clouds) build up into the stratosphere (in either middle or low altitudes) and that the thunderstorms of the world are actually a minor component of the overall meteorological machinery by which vapor moves up to the tropical stratosphere—most of it ascending slowly over an enormous area in the rising branch of circulation near the Equator. But in any event, the all-important point that the non-meteorologist should realize in connection with this is that we are already taking into account any and all such effects when we start hazard analyses (as in the skin cancer question below), with the *observed* average stratospheric water vapor content of about 5 ppm by volume.

(4) *SST pollution effects are unimportant, since they will constitute only about one per cent of the pollution from other technologies.* This argument has several fallacies in it. First, it takes as "SST pollution" only the kind of particulate and gaseous pollutions that tend to cause pollution difficulties in the lower atmosphere (sulfur dioxide, hydrocarbons, soots), and it quite casually ignores the point that, in the stratosphere even such a seemingly harmless exhaust product as water vapor can lead to serious disturbances. Secondly, it ignores the point stressed above: the roughly 100-fold greater holdup time (turnover time) of the stratosphere. Contrasted with the troposphere, where other present forms of atmospheric pollution are occurring, if SSTs put out only about 1 percent as much pollution as all other polluting technologies, then the 100-fold greater holdup time characteristic of the part of our atmosphere in which SSTs will cruise will just about cancel out that advantage. That is, storing up for a hundred times as long the emissions of an SST emitting only 1 percent as much pollution would bring the SST pollution-levels up to about par with those of all other technologies.

(5) *If SSTs are going to pollute our atmosphere, it is better to have it polluted by U.S. SSTs than by the same number of foreign-produced SSTs.* Sometimes it is suggested that it is pointless to debate the present SST decision because if we don't build SSTs, somebody else will. I believe a sounder

viewpoint is this: Whether we build and fly many hundreds of stratospheric commercial transports or whether some other country does, exactly the same careful scientific assessment of potential global (or hemispheric) environmental hazards has to be conducted. Entirely independent of who builds high-powered SSTs that will contaminate the sensitive and reactive stratosphere, the hazard burdens will be much the same, since stratospheric winds will spread the effects over most of the northern hemisphere. The real question at stake is thus the question of whether it is acceptable *to any and all nations* to have operating in the stratosphere a heavy air-transport technology which might impose globally unacceptable environmental burdens affecting any or all national interests. The inherently international characteristic of the problem, when properly appreciated, requires that the United States or Russia view the Concorde program just as critically as the British or Russians must view the American program. The basic question is whether the stratosphere is going to be just too sensitive to the kinds of transport technology now envisaged by aeronautical engineers who have made their impressive advances without fully examining environmental implications of the powerful and fast vehicles they have been designing with such skill.

(6) *The sensible and conclusive way to sort out all of these questions about SST environmental effects is to build several prototypes, fly them in the stratosphere, and make direct measurements to settle all of the uncertainties.* Although such a suggestion has a rather plausible ring, careful examination of where the really crucial environmental questions now lie shows rather conclusively that availability of a few flying SST prototypes will do almost nothing to settle these scientific controversies. There are far more scientific uncertainties than I shall be able to pinpoint here; but, with only a few exceptions, the kinds of research now needed to clarify these problems involve laboratory or computer work, or biological studies having no relation at all to the availability of prototype SSTs. Our difficulty is that we lack basic scientific understanding of a number of key questions raised by proposals to initiate high-altitude transport technology. As with all too many past examples, we have failed to conduct a broad and vigorous program of basic research, so that we just do not have all the answers at hand when technological change suddenly calls for assessing hazards. Flying models of the SST may be useful in checking engineering-feasibility questions but they will, unfortunately, be essentially useless in providing answers to most questions on environmental side-effects that have emerged from recent inquiries.

AN EXAMPLE OF AN OVERLOOKED SST ENVIRONMENTAL HAZARD: INCREASED INCIDENCE OF SKIN CANCER . . .

Past experience with new technologies offers a number of historical examples of the general principle that entirely unanticipated environmental difficulties, often of a rather subtle nature, may ultimately come to light after years of operation with that new technology (pesticide technologies, release of mercury from seemingly harmless industrial processes, radiation hazards, lead tetraethyl gasoline additives, and the like). Certainly it must be agreed that, in trying to avoid many more such experiences, we need to find ways of assessing new technologies far more carefully than we have done in the past. A National Academy of Sciences report on technology assessment has provided a penetrating analysis of these challenges, and has drawn particular attention to the need for detecting those potential side-effects of new technologies before the latter attain so advanced a state of development that too large an economic and social investment has been made to stop them.

It was very much in that spirit that I have re-examined certain earlier conclusions that ozone reduction due to water vapor added to the stratospheric ozone layer would be too small to have discernible biological importance. Although the full argument that has led me to the view that SSTs might cause adverse effects large enough to be of public health concern is too lengthy to be completely detailed here, I wish to outline its main features, since I believe that the argument is strong enough that it must now be carefully weighed into the present decisions on the SST program.

(1) *Carcinogenic effects of solar ultraviolet radiation.* That skin cancer, especially of the basal cell and squamous cell types, is caused chiefly by prolonged exposure to

solar ultraviolet radiations has now been attested in so many ways as to be essentially beyond dispute. Among the items of evidence supporting this conclusion, those of particular concern relative to SST effects include the following: *(a)* In all parts of the world there is a systematic tendency for higher skin cancer incidence in regions characterized by low average ozone amounts overhead, by high percentages of clear skies, and by short airpaths for incident solar rays. Briefly, all of these factors tend to imply higher UV (ultraviolet) dosage rates at lower latitudes, and skin cancer incidence is indeed found to increase correspondingly as one moves from higher to lower latitudes. *(b)* In any given area, skin cancer incidence is known to run much higher among persons who spend a great deal of time out of doors (farmers, ranchers, construction workers, sportsmen, seamen, and the like). *(c)* Longstanding clinical experience shows that skin cancer lesions appear predominantly on exposed portions of the body; about 85 to 90 percent of all lesions, in fact, occur on head and neck areas, which are least covered by clothing. *(d)* For the same reason, average incidence for males exceeds that of females in essentially all parts of the world. *(e)* Light-complexioned persons exhibit markedly higher skin cancer incidence than do dark-skinned persons; persons of Celtic derivation seem particularly vulnerable according to many studies, while Negroes exhibit much lower incidence than do Caucasians. *(f)* Albinos provide particularly dramatic evidence (albeit of very distinctly different nature) of vulnerability to solar induction of skin cancer. *(g)* Laboratory irradiation experiments demonstrate induction of skin cancer at wavelengths near 3000 angstroms (in the UV region). *(h)* Increasing incidence of skin cancer in recent years appears explainable in terms of changing recreational habits leading to greater average solar exposure and in terms of changing clothing habits, especially among women.

(2) *Critical role of the ozone layer in filtering solar UV.* Despite the seemingly small total amount of ozone in the stratosphere (equivalent in mass to a layer of sea-level air only about 3 millimeters thick), the strong absorptivity of ozone for wavelengths near 3000 angstroms and below serves to filter out much of the solar UV so that, for most purposes, it is accurate enough to say that the ozone produces a cutoff near 2900 angstroms. It has long been known that this filtration effect is of critical biological importance to all forms of life, especially animal life; but in just the past few years there has emerged a still more impressive and still more fascinating series of indications that stratospheric ozone has, in fact, been a crucial limiting factor throughout most of the evolutionary history of terrestrial life. The evidence is now mounting very rapidly that various lifeforms (ranging from microorganisms to humans) now survive in the face of existing solar UV exposures only because of having evolved astonishing and fascinating protective mechanisms or UV damage-repair mechanisms at the cell-biological level. I cannot here do justice to this impressive body of biological evidence; suffice it to say that DNA damage results from UV irradiation, and the more so the shorter the active UV wavelengths, since DNA absorption is a maximum near 2600 angstroms. Furthermore, mounting evidence implicates thymine dimerization in DNA as either a controlling or a contributing factor in UV induction of skin cancer, some of the most cogent evidence thereof having been turned up very recently in special studies dealing with pigmented skin tissues.

(3) *Implications of the north-south gradient in skin cancer incidence.* Various epidemiological studies have revealed a north-to-south increase in average skin cancer incidence in the United States that amounts to about an eight- to ten-fold higher incidence (measured in numbers of new cases detected per year per 100,000 total population, and ranging from around 25 per 100,000 per year averaged across the northern tier of states to around 200 to 250 per 100,000 for the southern tier of states). Strong corroboration of this kind of systematic latitudinal gradient of skin cancer comes from all parts of the world, and the gradient is marked enough that it can be detected in data from within a single large state like Texas, or within rather small countries like England or Japan.

This north-south gradient of incidence results from the interaction of a number of factors, including average annual number of hours of outdoor exposure, cloudiness, solar elevation angles, and total overhead concentrations of stratospheric ozone. For fairly obvious reasons, annual dosages are dominated by summertime exposure, and when data on cloudiness and effects of sun angle are examined, fairly small latitudinal effects in the U.S. are found. A larger gradient effect is brought in by length of time per year in which temperatures are warm enough to permit appreciable amounts of out-of-doors work and recreation. Without reviewing all details, let me say that considering all of these factors has led me to assign no more than half the total dosage gradient to factors other than ozone differences. (I suspect this may be underestimating the relative importance of the ozone gradients.) Taking that value along with reported medical data on cancer incidence yields a rough, but I believe meaningful, calibration figure of a variation of about six percent of skin cancer incidence for every one percent of variation of columnar total ozone overhead. Unfortunately, laboratory studies using experimental animals have never been carried out in a way permitting direct cross-check on this "amplification factor" of around six; but such data as are in the literature are at least not incompatible with such a factor. There is urgent need (on grounds broader than mere SST concerns) to secure far more such data in the near future.

(4) *Reductions of stratospheric ozone by chemical interaction with SST water vapor.* Beginning about five years ago, a series of investigations has revealed that naturally occurring stratospheric water vapor interacts with ozone to reduce its average concentrations by a substantial amount. Still more recently the theory of these photochemical interactions has been employed to estimate the percentage reduction of average stratospheric ozone that one might expect to result from SST operations. These predictions depend, of course, on the assumed numbers of SSTs and on certain atmospheric parameters, such as turnover time, mixing effects, and the like. Current estimates range from about two percent to about four percent ozone reduction.

Assuming an ultimate global SST fleet totaling the equivalent of 800 American SSTs (500 American SSTs plus the fuel equivalent of 300 more operated by foreign countries), using a figure of 6 hours/day in cruise-mode at or near 65,000 ft [and using other reasonable assumptions] . . . one obtains a predicted decrease in columnar total ozone of about four percent from the rise of water vapor concentration. However, again in the interests of conservatism (and adding to what I believe to be several other conservative biases built into other parts of my overall estimates), I have taken only a one percent reduction for purposes of the rest of the argument.

> *Editor's note:* The chemical reactions between SST exhaust emissions and the ozone layer in the atmosphere are complex. It is now believed that the major effect of ozone reduction is caused, not by water vapor, but by nitrogen oxides. The effect can be summarized in a simplified way by writing the following two reactions between nitrogen oxides (NO, NO_2) in the emissions and ozone (O_3) in the air, to make ordinary oxygen (O_2):
>
> $$NO + O_3 \rightarrow NO_2 + O_2 \qquad (1)$$
>
> $$NO_2 + O \rightarrow NO + O_2. \qquad (2)$$

(5) *Estimated increase of skin cancer incidence resulting from SST operations.* Considering just the United States, where the present annual skin cancer incidence now runs about 120,000 new cases per year, the foregoing figure of a one percent ozone decrease, together with the previously discussed six-fold amplification factor inferred from epidemiological data, would imply an SST effect on national skin cancer incidence amounting to perhaps 7000 new cases per year. If a four percent ozone reduction estimate were used, the corresponding estimated rise of skin cancer incidence would be about 30,000 new cases per year. If other conservative factors that I have used elsewhere in the argument were dropped, this figure might be doubled again. Here I prefer a round-number estimate near 10,000 new cases per year.

(6) *The amplification factor.* Because the literature on UV carcinogenesis was found to contain nothing resembling a well established mathematical model of skin cancer induction, and because the roughly six-fold amplification factor, which I obtained from considering epidemiology and related factors, is rather crucial to these estimates, I have devoted a good deal of effort to finding a physical and biological basis for understanding whether such an amplification effect can be understood. Without here elaborating the point in detail, let me merely re-

mark that there does appear to be an entirely plausible chain of reasoning, tied up with the marked non-linearity of absorption of the carcinogenic wavelengths, combined with the absorption properties of DNA. Briefly, I find that a six-fold amplification is just about what one should expect if the peak of the UV carcinogenesis were narrow and fell near 2950 angstroms. In reality, UV carcinogenesis almost certainly results from an appreciable range of wavelengths, but nothing in the biomedical literature is at all incompatible with an effective peak near the cited wavelength. This tends to give significant support to the overall argument.

(7) *Present conclusions on the SST skin cancer hazard.* I can fully understand why some persons might, on hearing that there was fear being expressed that increases of skin cancer could result from operating SSTs in the stratosphere, think that such a suggestion sounded ridiculous. But though there may well be errors in my analyses of the various parts of this problem, the prediction is far from being unsupported. The evidence is now quite strong that modest variations of stratospheric water vapor* concentrations could lead to just such modest ozone changes. And the evidence also is rather strong that modest reductions of stratospheric ozone would be reflected in increased average incidence rates for skin cancer. Finally, the purely biological and evolutionary evidence that we, as well as all other life forms, have evolved in ways leaving us only marginally protected from highly adverse effects of ultraviolet radiation is essentially incontrovertible. One needs, perhaps, to reflect on other examples of inadvertent modifications of our natural environment as a result of new technologies to be reminded that adverse effects have repeatedly unfolded as a consequence of causal chains that connect seemingly very distantly related events. In my own opinion, the present evidence points rather strongly to the conclusion that operating a major SST technology of the magnitude now under consideration here and abroad would, via the ozone-ultraviolet-carcinogenesis chain, lead to increased incidence of skin cancer on the order of ten thousand new cases per year within the United States alone. The world total would be somewhat greater, of course, but not greater by a large factor, since skin cancer is above all an affliction of Caucasians (rather than Negroes, Asians, or other pigmented ethnic stocks), so considerations of world geography and world climatology seem to imply that the brunt of the skin cancer burden, regardless of who builds and flies them in the Northern Hemisphere, would necessarily be borne by the Caucasian population of this country, with Europe bearing most of the remainder of the total burden.

My suggestion, based on analysis of just this one SST environmental hazard, is that this single side-effect poses by itself a difficulty of sufficient magnitude to postpone any immediate commitments which might tend toward subsequently irreversible pursuit of SST/HST high-altitude transport technologies. Until reliable answers can be obtained through appropriate research, I would have to suggest that these present estimates, albeit tentative, are much too disturbing to warrant further immediate moves toward SST/HST technologies. If careful research into all of the many factors underlying the above estimates should disclose that an estimate of some ten thousand new skin cancer cases per year is a gross overestimate, then this one of the several grounds for caution in the present SST decisions will disappear. My present surmise, however, is that it will probably be found to be on the low side.

*or, according to current thinking, nitrogen oxide. (Ed.)

20 ENVIRONMENTAL HAZARDS OF NUCLEAR WASTES

PHILIP P. MICKLIN

One of the more problematic ramifications of nuclear technology is the very complex task of safely managing long-lived, highly radioactive wastes. These are composed of certain of the radioactive [isotopes] produced in nuclear reactors by fission of the uranium or plutonium fuel. Krypton-85, cesium-137, and strontium-90 present the greatest management problems because of their long physical half-lives. Although plutonium and americium are not fission products, they and other *transuranic** elements produced in reactors constitute a small part of the wastes from reprocessed reactor fuels and also require long-term containment.

Because of their highly radioactive content, reactor wastes must be isolated from the *biosphere*† for centuries or longer. Consequently, the federal government has operated waste repositories since the latter part of World War II when plutonium production for weapons was begun. In recent years, private storage facilities licensed by the Atomic Energy Commission (AEC) have been established to handle the growing quantity of wastes generated by the reprocessing of fuels from nuclear power plants.

The greatest quantity of high-level wastes, approximately 85 million gallons, is contained at the three federal repositories: the Hanford Reservation near Richland, Washington; the National Reactor Test Site (NRTS) near Arco, Idaho, and the Savannah River Reservation near Barnwell, South Carolina.

In early 1973, 64 million gallons of waste were stored at Hanford in 151 underground tanks, ranging in size from 55,000 to 1 million gallons. Most of the tanks contain non-boiling wastes; only 20 are filled with high-heat, *self-boiling liquids*.‡ The non-boiling

*Elements heavier than uranium, such as neptunium, plutonium, and others. They do not occur in nature, but are either made artificially and used in nuclear reactors or are produced by the reactors as they operate. (Ed.)

†The portion of the earth on which life exists—the atmosphere, the soil and the fresh and salt waters. (Ed.)

‡Some solutions of radioactive materials emit radiation at such a high rate that the solution boils constantly. (Ed.)

wastes are stored in tanks where the liquid evaporates and a salt cake is formed. Strontium-90 and cesium-137 are first removed from the high-heat, self-boiling wastes, and then the low-heat residue is aged from three to five years before the residue is evaporated to salt cake. The separated strontium and cesium are [currently] being stored as liquid concentrate until facilities at Hanford to solidify, encapsulate and store these wastes under water will be operational.

Tank storage of liquid wastes, according to the AEC, "has proved both safe and practical". However, there have been serious problems at Hanford where liquid wastes have been confined the longest. Sixteen leaks, totaling nearly 350,000 gallons, have been found over the past 16 years. The most recent and largest accident took place from April 20 to June 8, 1973, when 115,000 gallons of low-heat waste—containing 40,000 *curies** of cesium-137, 14,000 curies of strontium-90, and 4 curies of plutonium-239—leaked. The leak occurred in a recently filled 29-year-old 533,000 gallon tank.

A 1968 report by the General Accounting Office (GAO) had warned that Hanford had a potentially serious situation in regard to the condition of its existing tanks, particularly those that had been in service for more than 10 years and had been weakened by the corrosive waste solution. The report noted the situation was especially grave because all the tanks were of a single-shell design—a steel liner encased in concrete—and thus provided no secondary containment if leakage should occur. It also said there was insufficient reserve storage to handle the wastes from one of the large, self-boiling waste tanks if it became necessary to empty it.

Improvements in storage practices have been instituted at Hanford. The solidification program has reduced liquid wastes in storage from 64 to 42 million gallons. Greater reserve capacity has been provided and formalized spare tank criteria have been developed. And most important, new double-shelled, million gallon tanks have been constructed to contain the most hazardous wastes. Nevertheless, a follow-up GAO review of waste management practices in late 1970 concluded that problems would continue to exist as long as large quantities of liquid wastes are maintained in old facilities. It also questioned the long-term efficacy of in-tank storage of solidified wastes and implied that reserve storage facilities were still insufficient.

*A curie is an amount of a radioactive substance—the amount which gives off radiation at the same rate as about one gram (one 28th of an ounce) of pure radium. (Ed.)

WHOM CAN YOU BELIEVE?

★ Tank storage has proved both safe and practical—AEC, 1969.

★ As long as large quantities of liquid wastes are maintained in old facilities, problems will continue to exist—GAO 1971.

★ At Hanford, 16 leaks, totaling nearly 350,000 gallons, have been found over the past 15 years—Micklin.

The other federal waste repositories have much smaller volumes in storage. By 1972, 4 million gallons of acid, high-heat aqueous waste had accumulated at the National Reactor Test Site (NRTS); and at the Savannah River facility, 18 million gallons of waste had been accumulated by 1973. Storage at Savannah River consists of 22 double-shelled, cooled tanks for high-heat wastes and eight uncooled tanks for low-heat wastes. Since 1960, Savannah River has had a program of in-tank evaporation of low-heat wastes to solids. High-heat wastes, to the extent commensurate with the structural and heat dissipation capabilities of the tanks, are also evaporated to a mixture of salt crystals and sludge. Waste produced at the NRTS facility is stored in cooled, underground, double-shelled tanks for a few years to age. Then it is *calcined** to a granular solid with a volume about one-tenth of the original liquid. The granular material is stored underground in cooled stainless steel bins inside concrete vaults. By 1973, more than half of the liquid wastes ever produced at the facility had been converted to solids.

Storage problems so far have been minor at the Savannah River facility. Four tanks have leaked 700 gallons of wastes but in only one instance did radioactive material contaminate the ground. In both the 1968 GAO study and follow-up review, concern

*Heated in an oven or furnace, leaving a dry residue. (Ed.)

was expressed about insufficient reserve storage capacity in case of the need to transfer wastes at the Savannah River site. The Idaho facility evidently has had no storage problems of note.

Although not presently posing a serious storage problem, high-level radioactive wastes from the processing of spent commerical reactor fuel are expected to grow rapidly and could reach 93 million gallons by the year 2000. Solidification would considerably reduce the volume to be stored but would not diminish radioactivity.

Computer calculations carried out by the author indicate that the accumulation of three key fission products, strontium-90, cesium-137 and krypton-85, from fuel reprocessing could reach nearly 45 billion curies by the end of the century. Most of the radioactivity will be in the form of stored wastes. However, the gas krypton-85 is routinely released to the atmosphere, although it may be collected and stored in the future. In any case, the amount of wastes accumulated by the year 2000 will have to be stored for nearly 500 years to allow for decay of the accumulated strontium-90 and cesium-137 to a low enough level (approximately one curie per cubic foot of solid material) that release to the environment would pose no serious threat. And to allow for the entrained transuranics, such as plutonium-239, to decay to innocuous levels would require storage for several hundred thousand years longer.

COMMERCIAL WASTE

More serious management problems are posed by *commercial wastes** than by federally generated wastes. First, because of a longer period of fuel irradiation, commercial wastes can attain heat and radioactivity levels six times greater per unit of volume. Thus they exert considerably more thermal and radioactive stress on their containment structures.

Second, fuel elements from commercial reactors must be transported long distances to reprocessing plants. Shipment, both by rail and truck, is in massive, cooled casks. A railroad cask may weigh 100 tons and contain three tons of irradiated fuel with a radioactive level of 15 million curies and a thermal generation of 70 kilowatts. The probability of accidental breaching and release of contained radioactivity is admittedly slight for an individual cask. But with annual shipments expected to reach more than 10,000 casks by the end of the century, serious accidents become a distinct possibility. The worst "credible" accident foreseen by the AEC, the release of seven fuel elements from a cask, would give a *100 percent lethal dose** of 1,000 *rem*† to a person standing 100 feet from the exposed rods for six minutes.

*That is, mostly from nuclear power reactors and processing plants. (Ed.)

WHOM CAN YOU BELIEVE?

⋆ Krypton-85 is routinely released to the atmosphere—Micklin.

⋆ The release of krypton-85 from reprocessing plants presents no serious long-term hazard—AEC, 1972.

⋆ Krypton-85 is a public health threat that should be eliminated by collection and long-term storage of the gas—EPA, 1972.

The first commercial facility, Nuclear Fuel Services Inc., a subsidiary of Getty Oil and Skelly Oil, in West Valley, New York, began processing fuel in 1966. But it has recently shut down for modifications and will not resume operations again until 1976 at the earliest. It had an initial capacity to handle 300 metric tons of spent fuel annually. In addition to processing fuel from commercial reactors, the plant has handled some AEC-generated wastes. In early 1973, 600,000 gallons of high-level liquid wastes were stored there. The waste is stored in a tank resting in a saucer, both of which are stainless steel, and enclosed in an underground concrete vault.

The long-lived, highly radioactive gas krypton-85 has been routinely released to the atmosphere. With the contemplated expansion of the facility to handle 900 metric tons of spent fuel annually, the plant could

*The dose of radiation which is lethal to 100 percent of a population. (Ed.)

†Rem is an acronym for Roentgen Equivalent Man; it is the unit of dose of ionizing radiation [that] results in the same biological effect as that due to one roentgen of x-rays.

release 9 million curies of krypton-85 to the environment each year. The AEC feels [that] the release of krypton-85 from reprocessing plants presents no serious long-term hazard. But the Environmental Protection Agency has taken the position that it is a public health threat that should be eliminated by collection and long-term storage of the gas.

A second fuel reprocessing facility is under construction near Morris, Illinois. The Midwest Fuel Recovery Plant (MFRP) is expected to begin operation in spring 1974. Built by General Electric, the Morris plant initially will process 300 metric tons of spent fuel annually with the capability to expand capacity to 500 tons. In contrast to the Nuclear Fuel Services' facility, the MFRP will immediately calcine its liquid wastes to a solid oxide to be welded inside stainless steel containers for underwater storage. But like Nuclear Fuel Services, the Midwest repositories will vent large amounts of krypton-85 to the atmosphere.*

Reprocessing and high-level waste storage facilities are normally located in sparsely inhabited areas with favorable hydrologic, meteorologic and geologic conditions as an extra measure of safety. However, the MFRP is an exception. It is located less than 50 miles from Chicago in a densely populated region, within a half a mile of the Kankakee River, and underlain by permeable *glacial drift*†. Accidental release of radioactive wastes would be more serious there than at other storage sites. The siting justification is that engineering safeguards have been taken to ensure against any such release.

One other commercial reprocessing facility, the Barnwell Nuclear Fuel Plant, is under construction. This plant, which is being built by Allied Chemical and Gulf Oil, is adjacent to the Federal Savannah River complex in South Carolina. Capacity will be 1,500 metric tons of fuel annually when operations are initiated in 1975. Initially, wastes will be stored as liquids in cooled stainless steel tanks placed in concrete vaults. After five years these will be solidified, encapsulated and placed in water-filled basins.

The above-mentioned plants should provide sufficient processing and storage capacity through the present decade. But many more facilities will be needed to handle the expected waste load during the 1980s and 1990s.

*As of November 1974, the MFRP plant has been built. Because of unsolved technical problems, however, General Electric does not intend to operate it. (Ed.)

†Rocks, gravel, sand, etc. deposited by a glacier. (Ed.)

FUTURE WASTE MANAGEMENT

The AEC considers that the tank storage of high-level wastes has proven satisfactory so far. But they admit this is not a long-term solution since careful management cannot be guaranteed over the centuries required for the wastes to dissipate their radioactivity. Furthermore, over extended periods natural phenomena, such as earthquakes, or human folly, such as war or sabotage, may result in the breaching of these containers. Consequently, the AEC has been investigating long-term storage modes with the hope that a scheme of permanent disposal can be found ensuring the isolation of wastes from the biosphere for hundreds of thousands of years and not dependent on human management.

The AEC is approaching the long-term management of commercial and federal wastes separately. According to a 1970 directive, commercial wastes are to be converted from liquid to solid within five years of their generaton, stored at the reprocessing plants for no longer than 10 years, and then sent to a federal storage facility. A decision as to the latter is still pending. Shipping these wastes to the envisioned federal management site could cost $2 billion. Hence, the preferred solution is permanent *in situ* storage. . . .

[Underground] salt formations are preferred for permanent disposal of commercial high-level radioactive wastes, since they are dry and impervious to water. They also have good thermal conductivity . . . , seal fractures rapidly by plastic flow, have good compressive strength, and are generally located in areas of low seismic activity. Many experts contend that once sealed in these structures wastes should be isolated from the biosphere for hundreds of thousands of years. Major salt deposits underlie some 400,000 square miles of the United States (see figure). However, those best suited for waste disposal—lying within 2,000 feet of the surface and with a minimum thickness of 200 feet—are located in Michigan, New York and Kansas.

The AEC has considered bedded salt for-

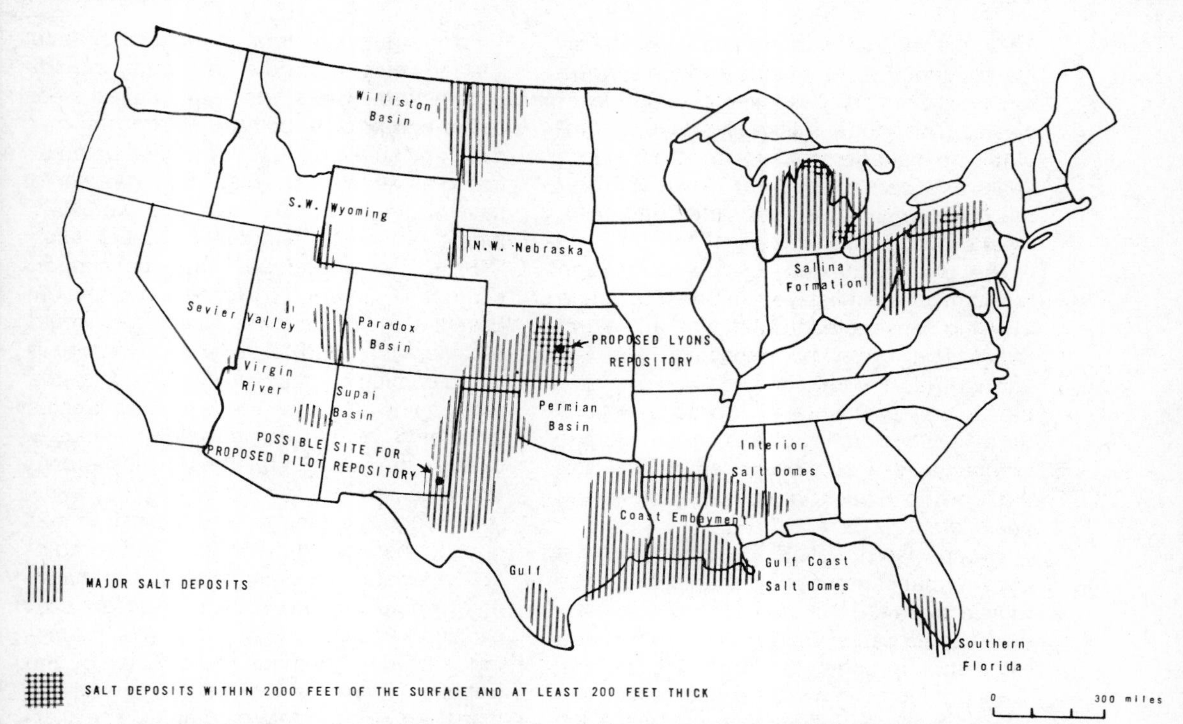

Major salt deposits and possible permanent high-level radioactive waste repositories in United States. [Source: F. K. Pittman, director, AEC Division of Waste Management and Transportation, Statement to FY 1974 Authorization Hearings Before Joint Committee on Atomic Energy, March 22, 1973, and Schneider et al., "Status of Solidification and Disposal of Highly Radioactive Liquid Wastes from Nuclear Power Plants in USA," in Proceedings of 1970 IAEA and AEC Symposium, *Environmental Aspects of Nuclear Power Stations* (Vienna: IAEA, 1971), pp. 369–386.]

mations as a storage mode since 1955 and initiated feasibility studies in 1959. During 1966 and 1967, a high-level solid waste disposal simulation was conducted in abandoned salt mines in central Kansas utilizing irradiated fuel assemblies and electric heaters to replicate storage conditions. The AEC considered Project Salt Vault a successful demonstration of salt storage and considered the beds in central Kansas as a suitable permanent repository. In 1970 it was announced that the abandoned Carey Salt Mine near Lyons, Kansas had been selected as a demonstration facility for waste emplacement. The impression was also given that, barring unforeseen occurrences, the National Radioactive Waste Repository would be established in central Kansas by the mid-1970s.

Subsequent events have been widely reported. The AEC's proposal met fierce resistance from local residents, the governor of Kansas, the state's congressional delegation and the Kansas Geological Survey. William Hambleton, Survey director, pointed out that the rock and salt underlying the Lyons area, owing to the presence of abandoned drill holes, shafts and mines, was akin to "swiss cheese," and that the distinct possibility existed that wastes buried in the salt formation would escape into groundwater. Because of the furor and the substantial evidence pointing to the unsuitability of the Lyons site, the AEC postponed its decision on the selection of a permanent repository.

The AEC still considers bedded salt formations in the Permian Basin most suitable for waste disposal. Currently, investigations are under way in other parts of Kansas and in southeastern New Mexico to locate favorable sites. A pilot waste facility will be constructed at a selected location before the end of the decade and 500 to 1,000 retrievable waste canisters will be stored. If demonstrated safe, it will be converted into an operating repository intended to handle all commercial wastes generated in the foreseeable future.

Because of the difficulties encountered in establishing a permanent waste repository in salt, the AEC has returned its attention to the concept of interim storage. On May 18,

1972, the AEC announced that it would use the technique of retrievable surface storage to manage solidified commercial high-level wastes. This mode, it is contended, will provide safe management for at least a hundred years and perhaps several centuries. Over this period, suitable permanent disposal means can be established.

The primary design for interim storage would involve a massive reinforced concrete building constructed to withstand all natural forces that might be expected, such as earthquakes and tornadoes as well as credible adverse accidental or intentional actions by man. Construction of this interim storage federal facility would be started in 1979, and the repository would go into service several years later.

Encapsulated, solidified wastes from the commercial reprocessing plants would be moved by rail to the federal facility in massive, cooled casks similar to those employed for shipping spent fuel from the reactor. Transporting solid wastes presents less of a hazard than shipping spent fuel, since the radioactivity and heat levels are lower, the wastes are contained in sealed cylinders, and there would be fewer than 10 percent as many shipments per year. The first few canisters are expected to be delivered in 1983 and approximately 80,000 canisters, representing the waste from nuclear fuel processed through the year 2000, could be in storage by 2010. Incoming canisters would be stored in water-filled basins for heat removal and radiation shielding. The water would be circulated through a heat exchanger for cooling.

The AEC maintains that this facility will be constructed to be totally safe and reliable. Nevertheless, as is the case with tank storage, heavy reliance will be placed on engineered safeguards. For example, loss of canister cooling could result in cylinder meltdown and release of wastes into the storage basins and possibly into the environment. The AEC considers this an "incredible" accident because of backup power and cooling systems, structural strength to withstand natural forces, and the period of time (approximately eight or nine days) it would take the cylinders to begin melting, during which corrective action could be taken. But over the expected life of this facility, an "incredible" series of natural and human circumstances might combine to produce just such an event.

An intriguing alternative form of surface storage being considered is sealing individual waste canisters into heavy-walled steel casks that provide both a heat transfer medium and shielding to reduce external radiation to an acceptable level. The casks would have walls 16 inches thick and weigh 35 tons; their surface temperature would reach 275°F. They would be stored on support saddles in the open air for heat dissipation. Essentially indestructible, such units would eliminate the need for reliance on sophisticated cooling and backup systems. The area required to store all the solidified wastes generated by the nuclear power industry through the year 2000 in this manner would be 700 acres.

Although the emphasis is on salt formations for permanent disposal of highly radioactive wastes, other more exotic management schemes have been proposed both within and outside the AEC. The AEC is funding a study of various atlernatives by the Battelle Northwest Laboratory, one of the managers of the Hanford complex. The alternatives being assessed are disposal on Earth, disposal in space and conversion of wastes to more easily handled forms by nuclear transmutation—that is, fissioning the long-lived wastes into much shorter lived elements whose radioactivity would diminish more rapidly.

Possible modes of Earth disposal besides salt include burial deep within other geologic formations, on the seabed and within continental ice sheets. Deep geologic emplacement of wastes, as much as 10 miles beneath the Earth's surface, would probably isolate them from water or mineral resources indefinitely. However, present drilling technology is limited to depths of five miles. Emplacement of waste canisters in deep ocean trenches has also been proposed. Theoretically, the wastes would be carried down into the Earth's crust. . . . Obviously, the magnitude and character of forces involved . . . have to be much more thoroughly understood before this proposal can be seriously contemplated.

A scheme for burial within continental ice sheets was detailed in the *Bulletin* [*of the Atomic Scientists* for January, 1973]. According to that proposal, an international repository would be established in Antarctica. Waste canisters would be transported to the continental interior, emplanted in the ice, and melt slowly through to bedrock. The authors' preliminary calculations indicated the wastes would be isolated from the biosphere for more than 250,000 years. A major drawback would be the long-distance transportation of high-level wastes, because of the possibility that ships carrying radioactive materials would sink and subsequently release huge amounts of radioactivity to the oceans.

Space disposal and transmutation are means for coping with the long-lived transuranic elements, such as plutonium, found in small amounts in the waste stream from fuel reprocessing. If these could be removed or altered, the containment time for waste would be shortened from hundreds of thousands of years to less than a thousand years. Sending transuranics into space would first require their separation from the waste mixture, and this is still not technically possible. Also, the launch vehicle would have to have an almost perfect reliability. The feasibility of transmutation depends on the development of the technology to segregate the transuranics and on [other] breakthroughs. . . .

In spite of unsolved problems, the AEC is convinced that long-term management of highly radioactive wastes poses no environmental hazards. They contend [that] they, or some successor organization, will be able to provide the careful control of these materials required into the distant future or until a reliable permanent disposal system is found. On this assumption, no threat is admitted in the expected rapid buildup of wastes from power reactors over the next several decades.

The AEC may be right. But if wrong, the consequences, borne primarily by future generations, will be catastrophic. As management and engineering systems are devised by humans, no matter how carefully designed they are inherently subject to failure, particularly over the long term. Hence, it may be prudent to consider stopping, or at least slowing, the head-long expansion of nuclear power and devoting our attention to developing less hazardous alternative energy sources until more complete answers are provided to this as well as other problems of nuclear technology.*

---

*References in the original article have been deleted. (Ed.)

PROGNOSIS

Kevin P. Shea

Predicting the flow of future events is no longer the exclusive province of seers and politicians. There is now a large cadre of technicians busily assembling information to be fed into computers in various combinations, in the hope that we might get a glimpse of what lies ahead for technological man.

The results of some of these computerized prognostications, filled as they are with gloomy statistics, have made headlines around the world, and have caused many international meetings to be convened for the purpose of developing strategies for avoiding the unhappy events predicted on the computer print-outs.

But one need not understand the intricacies of sophisticated computer models in order to recognize the hazards of continuing with business as usual for the next few decades. Indeed, some of those hazards have already been recognized, and a few measures have already been taken to ameliorate some of those technologies most offensive to the environment. DDT, for instance, has been banned from interstate commerce in the United States, a step that virtually assures its demise in the whole country. And legislative restrictions have been placed on some other highly toxic pesticides and on the PCBs we read about in Section Two. While these actions are laudable, they are piecemeal to be sure, and most have been undertaken only after serious environmental damage had already been done. Furthermore, actions such as these are limited. The problems we are dealing with transcend the jurisdiction of a single county, state or nation. They are global problems to which all countries, rich and poor, are exposed. While it is the industrialized nations that contribute most heavily to the situation, it is almost certain that the developing nations, left behind as they were by the technological revolution and eager to catch up, will soon be using some of the same environmentally disruptive but economically exploitive technologies that have been developed in the more advanced nations. This is certainly an understandable situation, for the short-term priorities of a developing nation—food, housing and employment—are far more urgent than long-term consideration of the environment.

An outstanding example of the exportation of modern technology, with all its attendant problems, is embodied in the multitude of agricultural programs known collectively as the "green revolution." In essence, the green revolution is the wholesale transfer of highly technical agricultural production techniques to underdeveloped countries for the purpose of increasing their yields of essential grain crops. The programs center on the use of specially-developed, high-yielding varieties of grain that can be grown only with the help of the intensive use of fertilizers, heavy application of pesticides, and specialized equipment. While these programs have indeed proved that they can increase production in the short run, their usefulness in the long run is questionable. Not only must they inevitably result in serious environmental disruption, but the heavy capital investment required to initiate and maintain them is almost certain to result in a variety of sociological dislocations. In reviewing the events that took place in the Canete Valley in Peru (Article 18), Doutt and Smith have already given us some idea of what can happen in an underdeveloped country when modern technology is applied without regard to existing local conditions.

We can also expect a proliferation of western technology in the field of nuclear energy, and therefore a worsening of the problems of nuclear wastes, as detailed by Philip Micklin in Article 20. The recent petroleum shortage and the skyrocketing world prices for crude oil will almost certainly give impetus to the spread of nuclear technology as an alternate source of energy in the emerging nations, and this in spite of the fact that the most serious environmental problem of nuclear technology, the storage of those highly poisonous waste products, is yet to be solved.

In agriculture, in energy and in many other ways, then, we can expect the underdeveloped nations in an age of rising expectations to make a strong effort to bring to their peoples a share of the wealth heretofore enjoyed only in the industrialized nations of the world. And with this effort will come an increase in the scope of the world's environmental problems, complicated by increasingly complex political interactions. The political problem—how nations can cooperate with one another to deal effectively with international pollution problems—is something that cannot be computerized. National goals are far too diverse and far too sensitive to change for anyone to attempt a rational prediction in this area. Nevertheless, in the coming decades international cooperation is likely to be the most important arena in which the trend of environmental degradation might be reversed.

But even the most pessimistic environmentalist must admit that, however dim the future looks, there are some mildly reassuring signs. While the recent recognition of an "energy crisis" was a wrenching experience for almost all the people of the world, it did stimulate the consideration of more rational and less polluting technologies for producing energy. It also clearly demonstrated, at least in the United States, that the growth of demand for petroleum products is not inevitable, and that the growth projections which the energy industries have traditionally been presenting to the public may well have been greatly exaggerated.

In agriculture as in energy production, there certainly exist better ways of doing things than we are now using. Given the pre-eminence of agriculture in the catalogue of human enterprises, it is surprising that the innovative ideas described by Wolff and by Doutt and Smith (in Articles 17 and 18) have not been utilized more extensively. But perhaps their time has now come.

An even more reassuring sign is that the public is beginning to take an interest in the doings of technology, and is beginning to insist that large-scale technical programs which entail obvious direct or indirect effects on the general population be fully discussed in public before they are undertaken. The public outcry that killed the SST program (probably more on economic than on environmental grounds, however) may be just the beginning. In the view of a large segment of the general public, scientists have always hidden behind a wall of confusing jargon and a self-righteous image as truth seekers. This wall is now gradually coming down, and we are beginning to realize that scientists have no special expertise in making the social and political judgments that are so often made necessary by the technological applications of their work. A strengthening of that realization more than any other is probably our greatest defense against further abuses to our earth's life-supporting systems.

Human embryo in an artificial womb. (Fritz Goro, Time Magazine.)

SECTION FOUR

SCIENCE, LIFE AND LOVE

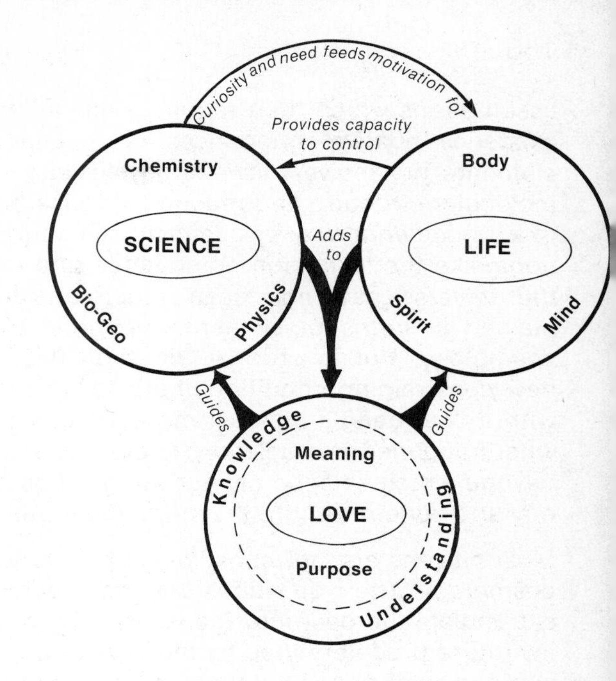

The Problem

Since time immemorial, man has asked the questions, Where do we come from? What are we? Where are we going?

And for thousands of years his religions have provided the answers, assigning him a position in the universe and giving meaning and guidance to his life. But today, at least in western European and American culture, the great images of the dominant religious tradition, the Judeo-Christian, seem to have lost their power over the minds of a large number of people, if not most of them. Yet no alternative view of man seems to have taken over the life-enhancing function performed by the Judeo-Christian view of God, of the world, of society, and of man.

According to a vague opinion held by many people, the world-view called "science" has provided this alternative by superseding our religious beliefs. But this opinion, as we shall see, is neither well founded nor well understood, nor is it generally accepted. In this Section we will examine the successes and failures of science in providing answers to these three basic questions, which are central to our everyday being.

Can science explain the origin of life, and of man? Can what we experience as love, for example, be reduced to biochemical reactions? Can the riddles of human thought and consciousness be understood in terms of the chemistry and physiology of the brain? Is our future as human and social beings determined or conditioned by the laws of physics and chemistry? And, ultimately, can man in the substance of his humanity be understood in terms of molecular structure and the kinetics of chemical reactions?

The answers to these questions are, in principle, generally accepted by experts to be partly affirmative and partly negative. Yet they are widely misunderstood by the public. For example, the origin of life can indeed be explained with a very high degree of probability in terms of chemical reactions, but in order to do so we have to make a variety of additional assumptions which then require explanations in themselves. Neither the existence, relative abundances, nor spatial distributions of the chemical elements, nor the very laws of physics and chemistry that determine both molecular structure and reaction kinetics, for example, can be explained in a fundamental way by chemistry. Of course, some of these presuppositions, like the formation, abundance and distribution of the elements in the universe, have now been rather satisfactorily—in broad outline—explained in terms of elementary-particle physics, nuclear physics, and cosmology. But in order to do even this, additional physical laws and new constraining conditions have to be assumed. If the universe began with a "big bang", as most modern cosmologists believe, we might ask what happened immediately before? There appears always to be a limit beyond which science cannot go, even as it grows more and more general and penetrates into the more and more distant past.

Most citizens are not consciously concerned, however, about their own cosmology. Their questions are more personal and immediate: Can we extrapolate the past into the future? To what extent are my present and my future predetermined by the laws of physics and chemistry? Does the world-view of physics and chemistry include ethical values to guide us in making the decisions that affect the future of science, of technology, and of society?

Indeed, the existential questions to be asked of modern science come even closer to us as persons when they ask not only about life, but about love. What indeed *is* "love"? Among the ideals of western culture, that which stands out far above all others is "love", yet the word is used, and misused, in a hundred different meanings. Can science make us more loving? Can loving determine what science we should or should not pursue? And what about dying? Is death the completion of, the end of, life and loving? In what conceivable way is science involved in these very basic human questions of life, love, and death?

We believe that our readers are interested in questions such as these. Accordingly, we have arranged the reading selections of this Section of the book into two groups, each of which treats one side of the interaction between "cold science" and human values: *Science's Control of Life* and *Humanity's Control of Science.*

FIRST PROBLEM: SCIENCE'S CONTROL OF LIFE

The first five selections (Articles 21–25) are concerned with man's capacity to modify himself. To what extent can "genetic engineering" change the species into a more desirable shape? Can man alter his body, his mind or his spirit by chemical manipulations? And if he can, *should* he? We know well that the most important public perception of science's benefits is in medicine—benefits which are in no small measure the fruits of the basic science of chemistry. But what of the extensions of medicine's new-found abilities into such fields as deliberate mind control by drugs or by electrical currents? The ethical implications of these questions are certainly in need of very careful scrutiny.

Almost daily, we read in newspapers and magazines about explorations into the origin of the universe, of our planet, and of life itself. These are modern scientific attempts to replace the age-old feeling that man is the center and the epitome of universal creation. It may be high time now for us to come to a balance between a too self-centered view of cosmology and some of the recently fashionable views which suggest that man is *merely* another animal and the earth *merely* one among billions of planets where even more highly developed forms of life may abound. Man is certainly set in and is fully a part of a vast, complex natural order, yet he is easily distinguished from all other living species by his consciousness of the laws of nature and by the extent to which he transforms his own environment. The relationship of man to nature may perhaps best be described by the word "steward": a responsible, thoughtful manager.

Article 21 is taken from the report of a special committee of the National Academy of Sciences, the most prestigious body of scientists in the United States. This special committee, formed of National Academy members and chaired by the Academy's president, Dr. Philip Handler, was formed to assess the relationship of biological research to the future of mankind. Its report shines a new light on what we mean when we speak of "man". The following excerpts from TIME magazine (Articles 22, 23 and 24) deal with the capacity man has developed, not only to control nature and his own environment, but also, through chemical and electronic means, to control his *body* and his *mind*.* As you read through these articles, you will note a progression: first, a discussion of what modern science *can* do to modify mankind; but then, the question, "Who decides what *should* or *should not* be done?" The last selection in this group (Article 25) therefore deals with the subtler questions involving man's capacity to intervene in his own future.

SECOND PROBLEM: HUMANITY'S CONTROL OF SCIENCE

While the readings in the first group illustrate some modifications of the human condition that are now within man's scientific grasp, they raise the question, "Should there be any controls over this kind of scientific work?" And if so, who is to decide which particular piece of research or which technological developments are desirable, or should even be permitted? And on what grounds?

The great French philosopher-lawyer-theologian, Jacques Ellul, felt a decade ago that our technological society was in trouble because we had permitted our technology to have a life of its own, to determine its own course, independent of the society that created it. Has man indeed loosed something he cannot stop, or are science and technology still merely man's tools, fully under his control?

Our second group of selections is four in number. The first (Article 26) is an excerpt from an address to the graduating class of the University of Kentucky College of Medicine—an address to the new doctors who are

---

*The reader who is interested in some of the basic biochemistry of the structure and function of DNA and RNA is referred to the same issue of TIME (19 April 1971), where an excellent treatment of this topic, with many clever illustrations, is to be found.

beginning their profession in a new kind of world in which they must face the questions we have just raised. The speaker, Dr. William N. Hubbard, Jr., was Dean of the University of Michigan School of Medicine at the time of this address, and is now President of the Upjohn Company.

In Article 27, Dennis Gabor, a Nobel laureate in physics, comments on what he calls "existential nausea": the widespread notion that all the products of science and technology have actually made life poorer, not richer in meaning. He emphasizes that in our complex technological world, excellence in knowledge must be matched by excellence in ethics, and he proposes an interesting ethical scale for measuring an individual's degree of responsibility to the common good.

Next, in Article 28, Harvey Brooks, Dean of Engineering and Applied Physics at Harvard and one of the principal architects of science policy in the United States, discusses the mutual interdependence of technological progress and social values.

Our final selection (Article 29) deals with the ultimate condition of man: death. Dael Wolfle of the University of Washington presents some of the ethical problems which arise from medical technology's ability to prolong life. Is it time, he asks, for a rethinking and refinement of our feelings about the quality of human life in its terminal phase, both on an individual and on a societal level?

Rustum Roy
G. R. Barsch

21 THE NATURE OF MAN; THE OPPORTUNITIES

THE COMMITTEE ON RESEARCH IN THE LIFE SCIENCES; PHILIP HANDLER, CHAIRMAN

The forces shaping the short-term future of man, perhaps to the turn of this century, are apparent and the events are in train. The shape of the world in the year 2000 and man's place therein will be determined by the manner in which organized humanity confronts several major challenges. If sufficiently successful, and mankind escapes the dark abysses of its own making, then truly will the future belong to man, the only product of biological evolution capable of controlling its own further destiny.

Social organizations, through their political leaders, will determine on peace or war, on the use of conventional or nuclear weapons, on the encouragement or discouragement of measures to limit the growth of populations, on the degree of increase in food production and on the conservation of a healthy environment or its continuing degradation. . . .

Man, a highly social being, is an animal as well. In form and function, development and growth, reproduction, aging, and death he is a biological entity who shares the attributes of physical life with the millions of plant and animal species to whom he is related. This relationship, known since prescientific times, became part of established science long before the theory of evolution was proposed. It is the reason why studies of fungi and mice, flies and rabbits, weeds, cats, and many other types of organisms have contributed to understanding man and to improving his health and biological well-being, and why future studies with experimental organisms will bear on man's own future.

Man's mental attributes form a superstructure which does not exist independent of his organismal construction. Human thought is based on the human brain; the brain—and, hence, the mind, the "self"—of each person is one of the derived, developed expressions of the genes which he inherited from his parents. Man's capacities are, thus, inextricably linked to his genes, whose molecular nature is now understood to a very high degree, but which [at present] lie outside his control. The

From a woodcut by Helen Siegl.

social creations of man—language, knowledge, culture, philosophy, society—have an existence of their own and are transmitted by social inheritance from generation to generation. But they depend for both their persistence and change on the genetic endowments of the biological human beings who are subjected to them while, at the same time, making them possible.

In these considerations, "man" should be taken as inclusive of all human beings on our planet. The brotherhood of all men is not only an ethical imperative, it is based on our common descent and on the magnitude of the shared genetic heritage.

The brain of man has not increased significantly in size since his Cro-Magnon ancestor, perhaps not for many millennia before. When one day man accepts responsibility for his acknowledged power to control his own genetic destiny, the choice [among] various plans must be based on value judgments. When he begins to use the power to control his own evolution, man must clearly understand and define the values toward whose realization he is to strive.

Man's view of himself has undergone many changes. From a unique position in the universe, the Copernican revolution reduced him to an inhabitant of one of many planets. From a unique position among organisms, the Darwinian revolution assigned him a place among the millions of other species which evolved from one another. Yet, *Homo sapiens* has overcome the limitations of his origin. He controls the vast energies of the atomic nucleus, moves across his planet at speeds barely below escape velocity, and can escape when he so wills. He communicates with his fellows at the speed of light, extends the powers of his brain with those of the digital computer, and influences the numbers and genetic constitution of virtually all other living species. Now he can guide his own evolution. In him, Nature has reached beyond the hard regularities of physical phenomena. *Homo sapiens,* the creation of Nature, has transcended her. From a product of circumstances, he has risen to responsibility. At last, he is Man. May he behave so!

22

CAN WE RE-SHAPE MAN?

THE STAFF OF *TIME* MAGAZINE

Reshaping life! People who can say that have never understood a thing about life—they have never felt its breath, its heartbeat—however much they have seen or done. They look on it as a lump of raw material that needs to be processed by them, to be ennobled by their touch. But life is never a material, a substance to be molded. If you want to know, life is the principle of self-renewal, it is constantly renewing and remaking and changing and transfiguring itself.
—Doctor Zhivago *by Boris Pasternak*

Perhaps it was simply a matter of chance, a random throw of the molecular dice. Perhaps some greater, transcendent force was at work in the earth's primeval seas. Yet from the moment of its miraculous genesis three billion years ago, life has been continually renewing and remaking itself, an evolutionary process that has led to the appearance of a unique creature quite unlike any of those before him. Thinking, feeling, striving, man is what Pierre Teilhard de Chardin called "the ascending arrow of the great biological synthesis."

Now, only some 35,000 years after the birth of modern man—a brief interval on the evolutionary time scale—the arrow is pointing in a dramatic new direction. Not only has man begun to unlock the most fundamental life processes, but he may soon be able to manipulate and alter them—curing such killer diseases as cancer, correcting the genetic defects that account for perhaps 50% of all human ailments, lessening the ravages of old age, expanding the prowess of his mind and body. Says Caltech's Robert Sinsheimer, one of the architects of the biological revolution: "For the first time in all time, a living creature understands its origin and can undertake to design its future."

To an extent, man has already altered himself and his planet. Scientists can only guess at the genetic toll from radioactive fallout, chemical contamination and other assaults on the environment. Even man's noblest impulses are apt to offend against nature. While improved medical care assures the survival and reproduction of those with genetically caused mental and physical defects, it also ensures that an increasingly larger percentage of the population will be heir to these illnesses in years to come. Geneticist Theodosius Dobzhansky succinctly expresses the ethical dilemma. "If we enable the weak and the deformed to live and to propagate their kind," he says, "we face the prospect of a genetic twilight. But if we let them die or suffer when we can save or help them, we face the certainty of a moral twilight." . . .

The biological revolution could make some of the choices easier. In the future, defective genes may be excised by pinpoint laser beams and replaced by viruses acting as man's genetic messengers in the body. Anguished man may also find his mental burdens lightened, as he turns to anti-aggression and knowledge pills, or learns to stimulate his brain's pleasure centers with electrodes. . . .

In the long history of evolution, 100 million species of plants and animals have inhabited the earth. Of these, 98% are now extinct, unable to survive the challenges of a changing environment. Man himself may face such a life-and-death test. Unlike his predecessors on the evolutionary ladder, he has the capability to meet it—and to fail it even more grandiosely than did creatures with lesser brains and imaginations.

23

THE BODY: FROM BABY HATCHERIES TO "XEROXING" HUMAN BEINGS

THE STAFF OF *TIME* MAGAZINE

The remarkable advances in molecular biology during the past two decades have given man an understanding of the basic processes that shape his life and have placed within the realm of possibility medical achievements undreamed of a scant few years ago. As more and more of the once-mysterious life forces within the cell are defined in the logical language of chemistry, the way is being opened not only for permanent cures of genetic diseases but also for drastic changes in man's genetic makeup. The acquisition of the power to eliminate genetic imperfections and engineer entirely new characteristics for humans is, for all of its promise, a frightening prospect for those who believe that man should not tamper with his inheritance. Yet even before the structure of DNA was defined and the genetic code broken, doctors had begun, mostly by trial and error, to develop techniques of genetic medicine.

Man today is heir to a host of inherited imperfections, ranging from diabetes to degenerative nerve disease. Each individual, geneticists have determined, carries between five and ten potentially harmful genes in his cells, and these flawed segments of DNA can be passed down to his progeny along with the messages that determine whether a child will have red hair or blue eyes.

Nature itself takes care of the worst genetic mistakes. One out of every 130 conceptions ends before the mother even realizes she is pregnant because the defective zygote, or fertilized egg, never attaches itself to the wall of the uterus. Fully 25% of all conceptions fail to reach an age at which they can survive outside the womb, and of these, at least a third have identifiable chromosomal abnormalities. Still, as many as five out of every 100 babies born have some genetic defect, and Nobel-Prizewinning Geneticist Joshua Lederberg believes the proportion would be even higher were it not for nature's own process of quality control.

The most obvious deformities result from

chromosomal abnormalities. Down's syndrome, or mongolism, which occurs once in every 600 births, is caused when one set of chromosomes occurs as a triplet rather than a pair. Hydrocephalus, or water on the brain, and polydactyly, the presence of extra fingers or toes, also result from faulty genes.

But the majority of genetic stigmas have somewhat more subtle symptoms and occur when defective genes fail to order the production of essential enzymes that trigger the body's biochemical reactions. Phenylketonuria (PKU) is caused by the absence of the enzyme necessary for the metabolism of the amino acid phenylalanine; as a result, toxins accumulate in the body and eventually cause convulsions and brain damage. Cystic fibrosis, which causes abnormal secretion by certain glands and respiratory-tract blockage that can lead to death by pneumonia, is the most common inborn error of metabolism; it is believed to be caused by a deficiency in a single gene.

Most people are unaware that they are carrying defective genes until they have a deformed, diseased or mentally retarded child. While medical science has not yet developed the techniques for repairing the bad genes, it can increasingly determine that they are present. Genetic counselors can thus advise prospective parents on the possibilities that their offspring will be born with genetic diseases. Properly informed, a couple that runs a high risk of producing a defective child may well decide to forgo having children.

If both parents carry genes for diabetes, for example, the chances are one in four that their children will inherit an increased risk for developing the disease. If either parent actually suffers from diabetes, the odds are even worse. Members of one large South Dakota family afflicted with a rare degenerative nerve disease have been advised, for example, that the odds are 50-50 that any children they have will suffer loss of balance and coordination and die, probably of pneumonia, by age 45 (TIME, Jan. 25).

Genetic counseling once relied more heavily on mathematics than medicine to predict the chance of hereditary handicaps. But it is now possible for doctors to identify and catalogue chromosomes. If there are certain chromosomal abnormalities, the prospective parents are informed that they will almost definitely produce deformed offspring. While this knowledge may take some of the mystery and romance out of procreation, it also eliminates much of the uncertainty. As one geneticist puts it, "There is nothing very romantic about a mongoloid child or a deformed body."

An even more important technique enables physicians to examine the cells of the unborn only months after conception and to determine with accuracy whether or not the

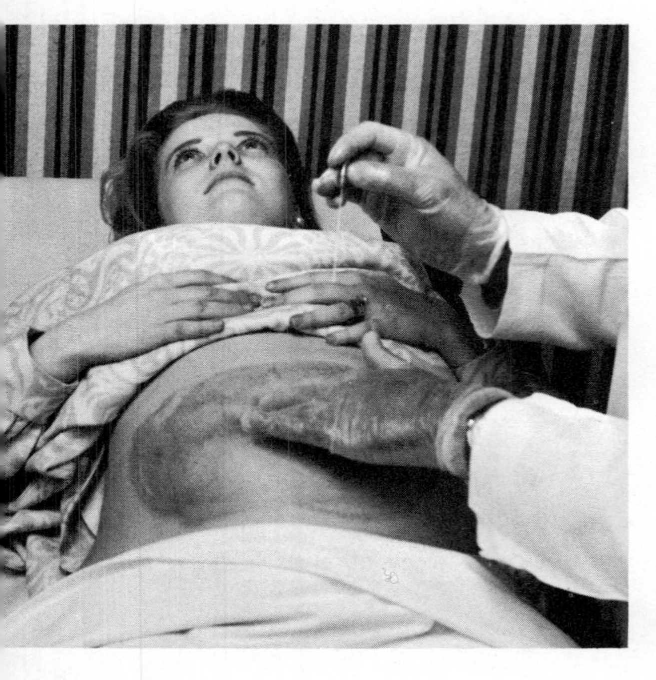

Diabetic woman undergoing amniocentesis. (From Leonard McCombe, *Time* Magazine.)

infant will inherit his parents' defective genes. The procedure is known as amniocentesis, from the Greek *amnion* (membrane) and *kentesis* (pricking); it is performed by inserting a long needle through the mother's abdomen and drawing off a small sample of the amniotic fluid, the amber liquid in which the fetus floats. Physicians then separate the fetal skin cells from the fluid and place the cells in a nutrient bath where they continue to divide and grow. By examining the cells microscopically and analyzing them chemically, the doctors can identify nearly 70 different genetic disorders, most of them serious.

Amniocentesis, performed between the 13th and 18th weeks of pregnancy, is not without some risk to both mother and baby. But in cases where family history leads them to suspect genetic defects, physicians feel that the benefits more than justify the danger; for the tests, which have been carried out on more than 10,000 women in the U.S. alone in the past 40 years, have proved extremely accurate. Using amniocentesis, Dr. Henry Nadler, a Northwestern University pediatrician, diagnosed mongolism in ten of 155 high-risk pregnancies tested. Subsequent examination of the fetuses showed that his diagnosis was correct in all cases.

At present, the woman who learns through amniocentesis that she is carrying a seriously deformed fetus has only two choices: abortion or the heartbreak of delivering a hopelessly defective infant. But the mother whose unborn baby is found to have one of several hereditary enzyme deficiencies has a more acceptable alternative, for medicine has developed techniques for treating many such illnesses. An amniotic test for fetal lung maturity, for example, has helped warn doctors when a child may be born with hyaline membrane disease, which blocks proper breathing. In those cases, birth can be delayed by sedation until tests show the baby ready to breathe on its own. Tests that permit prompt postnatal detection of PKU give doctors an opportunity to place babies so affected on special diets that prevent the accumulation of the deadly toxins and allow them to live relatively normal lives.

Some treatments are even possible before birth. Physicians routinely perform intrauterine transfusions on fetuses suffering from Rh disease, a genetic condition that results from the incompatibility of maternal and fetal blood.

Artificial insemination, once the exclusive province of livestock breeders, also offers escape from some genetic mishaps. An estimated 25,000 women whose husbands are either sterile or carry genetic flaws have been artificially inseminated in the U.S. each year, many of them with sperm provided by anonymous donors whose pedigrees have been carefully checked for hereditary defects. Some 10,000 children are born annually of such conceptions.

Doctors also see possibilities in artificial inovulation, a procedure in which an egg cell is taken directly from the ovaries, fertilized in a test tube and then reimplanted in the uterus. By carefully scrutinizing the developing embryo in the test tube, doctors could spot serious genetic deficiencies and decide not to reimplant it, thus avoiding an abortion later on. If the embryo is normal, it could even be reimplanted in the womb of a donor mother and carried to term there, enabling the woman either unable or unwilling to go through pregnancy to have children that were genetically her own.

Even test-tube babies, once the stuff of science fiction, are now not only possible, but probable. Dr. Landrum Shettles of Columbia University and Dr. Daniele Petrucci of Bologna, Italy, have shown that considerable growth is possible in test tubes. Shettles has kept fertilized ova growing for six days, the point at which they would normally attach themselves to the lining of the uterus. Petrucci kept a fertilized egg alive and growing for nearly two months.

Indeed, only development of an "artificial womb" capable of supporting life stands in the way of routine ectogenesis, or gestation outside the uterus, and now even this problem may yield to solution. Scientists at the National Heart Institute have developed a chamber containing a synthetic amniotic fluid and an oxygenator for fetal blood, and have managed to keep lamb fetuses alive in it for periods exceeding two days. Once their device is perfected, the baby hatchery of Aldous Huxley's *Brave New World* will be a reality and life without birth a problem rather than a prophecy.

Man may eventually be able to abandon sexual reproduction entirely. That startling and perhaps unwelcome possibility has been demonstrated by Dr. J. B. Gurdon of Britain's Oxford University. Taking an unfertilized egg cell from an African clawed frog, Gurdon destroyed its nucleus by ultraviolet radiation, replacing it with the nucleus of an intestinal cell from a tadpole of the same species. The egg, discovering that it had a full set of chromosomes, instead of the half set found in unfertilized eggs, responded by beginning to divide as if it had been normally fertilized. The result was a tadpole that was the genetic twin of the tadpole that provided the nucleus. Gurdon's experiment was also proof of what geneticists have long known: that all of the genetic information necessary to produce an organism is coded into the nucleus of every cell in that organism.

Man, say the scientists, could one day clone (from the Greek word for throng), or asexually reproduce himself, in the same way, creating thousands of virtually identical twins from a test tube full of cells carried through gestation by donor mothers or hatched in an artificial womb. Thus, the future could offer such phenomena as a police force cloned from the cells of J. Edgar Hoover, an invincible basketball team cloned from Lew Alcindor, or perhaps the colonization of the moon by astronauts cloned from a genetically sound specimen chosen by NASA officials. Using the same technique, a woman could even have a child cloned from one of her own cells. The child would inherit all its mother's characteristics including, of course, her sex.

Dramatic as cloning may be, it is overshadowed in significance by a technique that may well be practiced before the end of this century: genetic surgery, or correction of man's inherited imperfections at the level of the genes themselves. When molecular biologists learn to map the location of specific genes in human DNA strands, determine the genetic code of each and then create synthetic genes in the test tube, they will have the ability to perform genetic surgery.

A human embryo. (Photo courtesy of the Carnegie Institution of Washington.)

Some molecular biologists envisage using laser beams to slice through DNA molecules at desired points, burning out faulty genes. These would then be replaced by segments of DNA tailored in the test tube to emulate a properly functioning gene and introduced into the body as artificial—and beneficial—viruses.

The concept is not as farfetched as it sounds. Real viruses are merely segments of DNA (or RNA) surrounded by largely-protein sheaths; they penetrate the cell nucleus (leaving their sheaths behind) and take over the cellular DNA.

The potential of the technique is already being tested by an international research team in the treatment of two children whose hereditary inability to produce the enzyme arginase had resulted in severe mental retardation. The team infected the youngsters with a natural virus, the Shope papilloma, which contains DNA that triggers arginase synthesis. Although the experiment is expected to produce no improvement in the children's mental condition, it may belatedly trigger the production of the missing enzyme and prove that viruses can carry beneficial messages to the cells.

There is other evidence that the beginning of genetic surgery is not far off. Dr. Sol Spiegelman of Columbia University has synthesized an artificial virus that is indistinguishable from its natural model and has used it to infect bacteria and produce new viruses. He and his colleagues have little doubt that they will also eventually create "friendly" viruses and use them to cure disease rather than cause it—by using the viruses to stimulate the production of the chemical products upon which health and life itself depend.

Prophylaxis is important, but man's molecular manipulations need hardly be confined to the prevention and cure of disease. His understanding of the mechanisms of life opens the door to genetic engineering and control of the very process of evolution. DNA can now be created in the laboratory. Soon, man will be able to create man—and even superman.

Researchers have found that they can increase the life span of laboratory animals by underfeeding them and thus delaying maturation. This phenomenon, they believe, occurs because a smaller intake of food results in the formation of fewer cross linkages—connecting rods that link together and partly immobilize the long protein and nucleic acid molecules essential to life. If scientists can retard cross linking in man, they may well slow his aging process. Scientists also hope that they can some day do away with disease, genetically breeding out hereditary defects while breeding in new immunities to bacterial and other externally caused ailments. Finally, they look forward—in the distant future and with techniques far beyond any now conceived—to altering the very nature of their species with novel sets of laboratory-created genetic instructions.

Current predictions about the appearance of re-engineered man seem singularly uninspired. Some scientists argue that man's head should be made larger to accommodate an increased number of brain cells. They do not, however, explain what man would do with this additional gray matter; there is good reason to believe that man does not use all that he presently possesses. A few others note that the efficiency of man's hands could be increased by an extra thumb and his peripheral vision enhanced by protruding eyes—improvements that seem unnecessary in the light of man's expanding technology.

Some favor less obvious alterations. They have suggested that man be given the genes to produce a two-compartment stomach (a cow has four) that could digest cellulose; that mutation could be advantageous if man fails to increase his food supplies fast enough to feed the planet's growing population, but superfluous if he does. They also want man programmed to regenerate other organs, such as he now does with the liver, so that he can repair his damaged or diseased heart or lungs if necessary.

Others call for even more specialized humans to perform functions that in reality will probably be done better by machines. British Geneticist J. B. S. Haldane called for certain regressive mutations to enable man to survive in space, including legless astronauts who would take up less room in a space capsule and require less food and oxygen (larger and more powerful spacecraft would seem to be an easier and less monstrous solution). Haldane also suggested apelike men to explore the moon. "A gibbon," he said only half-jokingly, "is better preadapted than a man for life in a low gravitational field."

Eventually, scientists fantasize, man will escape entirely from his inefficient, puny body, replacing most of his physical being with durable hardware. The futuristic cyborg, or combination man and machine, will consist of a stationary, computerlike human brain, served by machines to fill its limited physical needs and act upon its commands.

Such evolutionary developments could well herald the birth of a new, more efficient, and perhaps even superior species. But would it be man?

24
THE MIND: FROM MEMORY PILLS TO ELECTRONIC PLEASURES BEYOND SEX

THE STAFF OF *TIME* MAGAZINE

In all of his 35,000-year history, Homo sapiens has found it harder to fathom the depths of his mind than to unlock the secrets of his body. But the discoveries of molecular biology may well show the way to a new comprehension; they may make it possible, through genetic engineering, surgery, drug therapy and electrical stimulation, to mold not only the body but also the mind.

Man cannot wait for natural selection to change him, some scientists warn, because the process is much too slow. Yale Physiologist José Delgado likens the human animal to the dinosaur: insufficiently intelligent to adapt to his changing environment. Caltech Biophysicist Robert Sinsheimer calls men "victims of emotional anachronisms, of internal drives essential to survival in a primitive past, but undesirable in a civilized state." Thus, by his own efforts, man must sharpen his intellect and curb his aboriginal urges, especially his aggressiveness.

To most laymen, the idea of remaking man's mind is unthinkable; "You can't change human nature," they insist. But many scientists are convinced that the mind can be altered because it is really matter. Explains Physicist Gerald Feinberg: "What sets us apart from inanimate matter is not that we are made of different stuff, or that different physical principles determine our workings. It is rather the greater complexity of our construction and the self-awareness that this makes possible."

That self-awareness resides in the brain, the organ about which scientists have the most to learn. To Physiologist Charles Sherrington, the brain's 10 billion nerve cells were like "an enchanted loom" with "millions of flashing shuttles." For some functions, M.I.T. Professor Hans-Lukas Teuber explains, brain cells are pre-programmed with "enormous specificity of configuration, chemistry and connection." Some are sensitive only to vertical lines, others only to horizontal or oblique ones. "Each of these little creatures does his thing," Teuber says.

In the hope of deciphering this staggering

variety, hundreds of scientists, including molecular biologists, in the U.S. and abroad, are now turning to brain research. One day in the distant future, their discoveries may help man to improve his already remarkable brain—for despite its dazzling versatility and subtlety, it is not without limitation. "Computers slashing from circuit to circuit in microseconds can cope with the input and response time of dozens of human brains simultaneously," Biophysicist Sinsheimer laments. Besides, the brain can call up only a limited amount of stored information at a time to focus it on a particular problem. And while it can grasp as many as 50 bits of visual information at once, it cannot file away more than 10 of them per second for later reference.

To most scientists, this reference system, or memory, is one of the most important tools of man's intelligence. Long before the development of molecular biology, Marcel Proust pondered the mystery of memory in *Remembrance of Things Past.* About a man's own past, he wrote that "it is a labor in vain to attempt to recapture it: all the efforts of our intellect must prove futile. The past is hidden somewhere beyond the reach of the intellect." In *Swann's Way,* it was a tea-soaked *petite madeleine* that touched off the hero's long-forgotten childhood memories. In the scientific world, the stimulus is sometimes a surgeon's probe. Montreal Surgeon Wilder Penfield, for example, while performing operations under local anesthesia, by chance found brain sites that when stimulated electrically led one patient to hear an old tune, another to recall an exciting childhood experience in vivid detail, and still another to relive the experience of bearing her baby. Penfield's findings led some scientists to believe that the brain has indelibly recorded every sensation it has ever received and to ask how the recording was made and preserved.

Initially, some brain researchers believed that memories were stored in electrical impulses. But scientists could not comprehend how a cranial electrical system, however complex its interconnections, could accommodate the estimated million billion pieces of information that a single brain collects in a lifetime.

Their doubts increased when they found that a trained animal generally remembered its skills despite attempts to disrupt its cerebral electrical activity by intense cold, drugs, shock or other stress; only short-term memory—of recently learned skills—was impaired. There was an obvious conclusion: while short-term memory may be partly electrical, long-term memory must be carried in something less ephemeral than an electric current.

That something, theorists believed, was chemical. Scientists had long known that chemical as well as electrical activity goes on in brain neurons: these cells carry on metabolism and protein synthesis like other body cells. Researchers soon learned that the leap of message-carrying nerve impulses across the gap between one cell and another takes place only with the help of chemical transmitter substances. One of these, acetylcholine, was promptly identified, and investigators began to look for other brain chemicals, specifically for varieties that might contain memories.

Their reasoning was that just as DNA carries genetic "memories," so other molecules might encode and carry information plucked from transient electrical impulses. Some early researchers proposed the idea of a separate brain molecule for each memory. The hypothesis of Swedish Neurobiologist Holger Hydén of the University of Göteborg was a bit more sophisticated; he thought that RNA was the key to memory formation and was encouraged in his belief by the results of his experiments with rats. When he taught them special tasks, he discovered that the RNA had not only increased in quantity but was different in quality from ordinary RNA. In short, what Hydén did was to lay the groundwork for a molecular theory of memory.

As Hydén's rat experiments demonstrated, RNA itself does not store memories; instead, it may play an intermediary role, stimulating the brain to produce proteins that are perhaps the actual repositories of memory. In one experiment inspired by that theory, University of Michigan Biochemist Bernard Agranoff taught goldfish to swim over a barrier, then injected them with puromycin, an antibiotic that prevents protein synthesis. When the injection was given hours after learning, it had no effect, suggesting that memory proteins had already formed. Injected just before or just after training, the drug prevented learning.

Other experiments based on the RNA-protein theory may demonstrate actual chemical memory transfer. Among the most publicized are those of University of Michigan

Psychologist James McConnell and Neurochemist Georges Ungar of the Baylor College of Medicine. McConnell works with planaria, or flatworms, conditioning them by electrical shock to contract when a light is flashed. He then grinds them up and feeds them to untrained worms. Once they have cannibalized their brothers, the worms learn to contract twice as fast as their predecessors. What may happen, McConnell theorizes, is that the first batch of worms form new RNA, which synthesizes new proteins containing the message that light is a signal to contract. Having consumed these memory proteins, the second group of planaria presumably do not need to manufacture so much of their own; they have swallowed memory, as it were.

Ungar's experiments are similar. Using shock, he conditions rats to shun the darkness they normally prefer, then makes a broth of their brains. This he injects into the abdominal cavities of mice, which seem to react with a parallel unnatural aversion to the dark. Moreover, the more broth Ungar injects, the faster the mice seem to learn this fear. His theory: the memory message (that darkness should be avoided) is encoded by the rats' DNA-RNA mechanism into an amino-acid chain called a peptide, a small protein that Ungar managed to isolate and then synthesize. His name for it: scotophobin, from the Greek words for "darkness" and "fear."

The experiments done by both men are hard to repeat, and investigators are still trying to decide whether the few apparent replications are sound. There is controversy, too, over the meaning of results: critics say it is hard to interpret the behavior of worms and other lower creatures objectively. Some say that Ungar may have discovered not a memory molecule but a molecule that blocked a normal response (to seek darkness) instead of teaching a new reaction (to seek light). Most investigators doubt that a single memory molecule will be found, but they believe that molecular biology will eventually reveal the secret of memory. If so, the blue-sky possibilities are limitless. It might be possible to develop "knowledge pills" that would impart instant skill in French, tennis, music or math. McConnell jokingly proposes another idea: "Why should we waste all the knowledge a distinguished professor has accumulated simply because he's reached retirement age?" His solution: the students eat the professor.

Many less frivolous proposals for improving memory and other aspects of mental life are emerging from molecular biology and genetics. It is known that genes do not cause behavior. But they influence it and set limits to physical structure, temperament, intelligence and special abilities.

Psychiatrist Alexander Thomas of New York University finds that babies show a characteristic style (easy, difficult or slow-to-warm-up) from their earliest days. While he admits that this temperament may develop in the months after birth, he does not rule out the possibility that it is inborn. Other life scientists warn that "when we strive for equality of opportunity, we must not deceive ourselves about equality of capacity." For example, it is believed that genetic influence is especially great in such areas as mathematics, music and maybe acrobatics. Unless genetic potential is tapped by the environment, it will not develop: kittens prevented from walking will not learn normal form and depth perception. Says Geneticist Joshua Lederberg: "There is no gene that can ensure the ideal development of a child's brain without reference to tender care and inspired teaching."

This interaction between environment and heredity is one of the factors that make it so difficult to change human characteristics. Another is that nearly all behavioral traits are polygenic—dependent on several genes. But even so complex a trait as intelligence may eventually come under the control of molecular biologists. Some scientists fantasize that super-geniuses will some day be produced by increasing brain size, through either genetic manipulation or through transplantation of brain cells to newborn infants or the fetus in the womb. (Such cells might be synthesized in the laboratory or developed by taking bits of easily accessible tissue from a contemporary Newton or Mozart and inducing them to turn into brain neurons.)

Another prospect is to alter genes so that babies will be born with rote knowledge—language skills, multiplication tables—just as birds apparently emerge from the egg with genetic programs that enable them to navigate. Some researchers hope to develop shared consciousness among several minds, thus pooling intellectual resources.

Most observers continue to feel that reining in man's aggressiveness is as important as spurring his intelligence. Harvard Neurosurgeon Vernon Mark advocates a non-genetic approach. "There are basic brain mechanisms that will stop violent behavior, and we are born with them," Mark asserts. To tap those mechanisms, scientists would like to develop an anti-aggression pill (estrogens, or female hormones, have already been used experimentally to inhibit aggressive behavior). Until they do, Mark and two Harvard colleagues—Psychiatrist Frank Ervin and Surgeon William Sweet—are fighting aggression by using surgery to destroy the damaged brain cells that sometimes cause violence in people with specific brain disease. Typical of their patients is a gifted epileptic engineer named Thomas, who used to erupt in rages so frenzied that he would hurl his children or his wife across the room. First, Mark and Ervin sent electric current into different parts of Thomas' brain; when the current sparked his rage, the doctors knew they had found the offending cells. Surgeons Mark and Sweet then destroyed them, and in the four years since, Thomas has had no violent episodes.

Physiologist Delgado has developed even more dramatic methods of aggression control in animals. In one famous experiment, he implanted electrodes in the brain of a bull bred for fierceness. Then, with only a small radio transmitter as protection, he entered the ring with the bull and stopped the angry animal in mid-charge by sending signals into what he believes was its violence-inhibiting center. Similarly, Neuroanatomist Carmine Clemente of U.C.L.A. has shocked cats into dropping rats they were about to kill. But neither man sees any early prospects for remote control of human aggression.

Other mental problems may well succumb to molecular biology. Many therapists resist the idea that emotional problems have biochemical equivalents; yet Freud himself believed that they do and that they would one day be identified. Researchers are already convinced that schizophrenia has some genetic basis, although, as Psychologist David Rosenthal explains, it is not the disease that is inherited but a tendency to it. As a match must be struck before it will burn, so must the tendency be triggered by something in the environment. No one is yet sure whether the trigger is cultural or familial, electrical or chemical, but some investigators back the chemical theory on the ground that certain drugs enable schizophrenics to live outside institutions, at least for short periods. To date, drugs for schizophrenia have been administered on a trial-and-error basis; as molecular biologists learn more, it will become possible to use specific drugs to achieve specific ends.

Further research may provide a bonus of new genetic, chemical and electronic ways to enhance sexual pleasure. Physicist John Taylor, in fact, professes to fear that sex will become so much fun that people will want to give up practically all nonsexual activities. Author Gordon Rattray Taylor predicts that it may become possible to "buy desire," or switch it on or off at will; the playboy might opt for continuous excitement and the astronaut for freedom from sexual urges during space flight.

Unlikely as it may seem, there are researchers who claim to have discovered something better than sex. At McGill University in Canada, Psychologist James Olds used electrodes to locate specific "pleasure centers" in the brains of rats, and then allowed the animals, electrodes still in place, to stimulate themselves by pressing a lever. Given a choice, the rats preferred this new pleasure to food, water and sex. Some pressed the lever as many as 8,000 times an hour for more than a day, stopping only when they fainted from fatigue.

Such experiments lead Herman Kahn of the Hudson Institute to predict that by the year 2000, people will be able to wear chest consoles with ten levers wired to the brain's pleasure centers. Fantasies Kahn: "Any two consenting adults might play their consoles together. Just imagine all the possible combinations: 'Have you ever tried ten and five together?' couples would ask. Or 'How about one and one?' But I don't think you should play your own console; that would be depraved."

Author Taylor, on the other hand, sees nothing wrong with solitary pleasure. Some day, he writes, a man may be able to put on a "stimulating cap" instead of a TV set, and savor a program of visual, auditory and other sensations. He and other futurists evision "experience centers" or "drug cafés" that would replace bars and coffeehouses. There, perhaps with the help of "dream machines," one might order a menu of "enhanced vision, sensory hallucinations and self-awareness." One might also be able to experience the mental states of a great man, or even of an animal. Molecular Biologist Leon Kass of the National Academy of Sciences projects a world in which man pursues only artificially induced sensation, a world in which the arts have died, books are no longer read, and human beings do not bother even to think or to govern themselves.

Some life scientists see even greater perils in man's new knowledge. "I would hate to see manipulation of genes for behavioral ends," warns Stanford Geneticist Seymour Kessler, "because as man's environment changes, and as man changes his environment, it is important to maintain flexibility." Professor Gerald McClearn of the Institute for Behaviorial Genetics at the University of Colorado agrees, explaining that a gene that is considered "bad" now might become necessary for survival in the event of drastic environmental change. "It is foolhardy to eliminate genetic variability," he says. "That is our evolutionary bankroll, and we dare not squander it. Species that ran out of variability ran out of life."

Such worries are probably premature. To some experts, the more radical forms of behavior control, especially genetic modification, belong to the realm of science fiction. Yet others believe that biological predictions are always too conservative, and that man will soon proceed, and succeed, with his experiments. If he does, he must prepare himself for a social and moral revolution that would affect some of his most cherished institutions, including religion, marriage and the family. With such possibilities in mind, Nobelist George Beadle has warned that "man knows enough but is not yet wise enough to make man."

25

THE SPIRIT: WHO WILL MAKE THE CHOICES OF LIFE AND DEATH?

THE STAFF OF *TIME* MAGAZINE

The quantum leap in man's abilities to reshape himself evokes a sense of uneasiness, a memory of Eden. Eat of the forbidden fruit, God warns, and "you shall surely die." Eat, promises the serpent, and "you shall be like God."

That temptation—to be "like God"—is at the root of the ethical dilemmas posed by molecular biology. In one sense, the new findings have continued the work of Newton, Darwin and Freud, reducing men to even tinier cogs in a mechanistic universe. At the same time, it was man himself who deciphered the code of life and who can now, in Teilhard de Chardin's phrase, "seize the tiller of the world." If he is only a bundle of DNA-directed cells, more sophisticated but hardly dissimilar from those of animals and plants, he can at least use that knowledge to improve, even to re-create himself. But should he?

In his persuasive 1969 book *Come, Let Us Play God,* the late biophysicist Leroy Augenstein argued that man takes the role of God by default or design and has always done so. Ecologically, he changes the very face of the earth: first with plows, then with dams, insecticides and pollution, he has seriously upset the balance of nature. His humane instincts and scientific curiosity team up to preserve life so well that the world faces a population crisis. Moreover, by extending the lives of those with defective genes, science increases the chance that damaging genes will be passed down to ever-larger portions of succeeding generations. Germany's pre-eminent Protestant ethicist, Helmut Thielicke, notes that men must recognize how "the act of compassion to one generation can be an act of oppression to the next." Thielicke argues that men must be willing to make hard choices. If society intervenes to keep alive the hereditarily ill (as he believes it should), then it must also be willing to intervene again, perhaps even sterilizing some with hereditary diseases.

This is only one kind of ethical problem raised by the new genetics, and it is already close at hand. Other problems are still in the far future, but how the dilemmas of population control are handled will set important patterns for later issues.

Population pressures increase the likelihood of widespread government drives, or even coercion, to limit births. Couples who are warned by genetic counseling that they risk producing deformed offspring would face far greater pressure than they do now to avoid having children: those with defective genes could become, in effect, second-class citizens, a caste of genetic lepers.

One current example illustrates the problem. Amniocentesis can now quite accurately predict whether a fetus is mongoloid; women carrying such abnormal fetuses are now encouraged, where it is legal, to have abortions. Already a number of medical planners are pointing up the cost-effectiveness of abortion in those cases. Unless the birth rate of mongoloid children is reduced, their care by 1975 may well cost some $1.75 billion nationally.

Methodist Paul Ramsey, Professor of Religion at Princeton and one of the top Protestant ethicists in the U.S., protests the aborting of such abnormal fetuses as an unjustified taking of human life. But he does not think moral men can avoid the problems of population and genetic crises. Indeed, he urgently recommends that society develop an "ethics of genetic duty." The right to have children can become an obligation not to have them, Ramsey asserts; it is shocking to him that parents will refuse genetic counseling and take the "grave risk of having defective children rather than remain childless." Dead set as he is against abortion in all but the most serious cases, Ramsey would prefer to see one parent undergo voluntary sterilization. "Genetic imprudence," he says, "is gravely immoral."

To Ramsey and others, genetic surgery—repairing, replacing or suppressing a "sick" gene—could be profoundly moral. Depending on the defect, genetic surgery before or after birth could prevent abnormality, and also insure that it was not passed on. Moral Theologian Bernard Häring of Rome's Accademia Alfonsiana applauds basic remedial intervention as "corrective foresight."

But Häring is one among many, both scientists and ethicists, who find it considerably harder to justify "positive" genetic engineering, restructuring the genes to make the "perfect" man. The prospect suggests apocalyptic possibilities: M.I.T. Biologist Salvador Luria approaches it "with tremendous fear of its potential dangers." Biologist Joshua Lederberg of Stanford University disowns such Utopian aims as a proper goal for serious biology, and even doubts that techniques sophisticated enough to achieve them could be perfected in the near future. But the possibility nonetheless tantalizes: Who would decide what qualities to preserve, and by what standards? Even remedial genetic engineering could pose a distressing problem if it achieved the ability to remove "undesirable" behavior tendencies. Asks Thielicke: "Would one try to eradicate Faust's restlessness, Hamlet's indecision, King Lear's conscience, Romeo and Juliet's conflicts?"

Human cloning, the asexual reproduction of genetic carbon copies, raises similar questions. Who shall be cloned, and why? Great scientists? Composers? Statesmen? When Geneticist Hermann J. Muller first broached the idea of sperm banks in *Out of the Night (1935)*, he suggested Lenin as a sperm donor. In later editions, Lenin was conspicuously absent, replaced on Muller's list by Leonardo da Vinci, Descartes, Pasteur, Lincoln and Einstein. Society could well be as fickle—or worse—about cloning. It might create a caste of subservient workers, as in *1984*, or a breed of super-warriors out of a "genetics race" between the U.S. and the U.S.S.R. An even more hideous nightmare would be the "clonal farm," where anyone could keep a deep-frozen identical twin on hand for organ transplants.

Such fanciful fears tend to obscure deeper ethical and practical objections to cloning. The process could be used, for example, to allow a woman to produce a child without passing on her own or her mate's defective gene. A cell nucleus from the genetically sound parent could be substituted for the nucleus in her egg. But even that quite reasonable application could introduce a novel set of complications. Would the cloned child develop a sibling rivalry with its biological parent? Would he face a severe identity crisis, being someone else's "duplicate"? Beyond such considerations, a number of scientists and ethicists would list cloning among those things that men should never do, even if they can. Says Embryologist Robert T. Francoeur, author of *Utopian*

Motherhood: "Xeroxing of people? It shouldn't be done in the labs, even once, with humans."

To many critics cloning is only one of several biological developments that threaten what Paul Ramsey calls "a basic form of humanity": the family. Ramsey thinks that artificial insemination by a donor, which is already fairly common, has opened the door to further invasions of family integrity. In his recent book *Fabricated Man,* he mentions other possible developments: artificial inovulation (the "prenatal" adoption of someone else's fertilized egg), "women hiring mercenaries to bear their children," and "babies produced in hatcheries." Beyond finding some of the possibilities repellent, Ramsey argues that they violate "convenant-fidelity," a bond of spiritual and physical faithfulness, between wife and husband or parent and child.

Francoeur, on the other hand, feels that the new embryology can lead to a fresh flexibility in the family structure. He favors host mothers (Ramsey's "mercenaries") because some women want children but cannot carry them to term. In an opposite way, artificial inovulation could be the means for a sterile mother to bear a child, even if not from her own egg. But he draws the line at artificial wombs, which, he says, "would produce nothing but psychological monsters." Others emphasize that the family itself must survive to fill important psychological needs. Molecular Biologist Leon Kass, who left the research labs to become executive secretary of the National Academy of Science's Committee on the Life Sciences and Social Policy, puts it effectively: "The family is rapidly becoming the only institution in an increasingly impersonal world where each person is loved not for what he does or makes, but simply because he is. Can our humanity survive its destruction?"

Beyond population control, beyond "Xeroxing" and patterning people, beyond the survival of the family lies the ultimate ethical question: the sanctity of life itself. The move toward new knowledge requires experimentation. The new generation of experiments, however, involves human life, and many moralists suggest that many of those experiments are intrinsically evil because they toy with life. They point, for example, to the experiments by Italian Biologist Daniele Petrucci, who in 1961 announced that he had kept a fertilized egg alive for 29 days *in vitro* (in the glass) before letting it die because it was monstrously deformed. Another Petrucci embryo lived for 59 days before it died because of a laboratory mistake. The Vatican, which sternly forbids all experimentation with fertilized eggs, demanded that Petrucci cease his investigations. He agreed to comply.

In a recent experiment conducted by Landrum Shettles at Columbia University, a 100-cell human embryo growing in a petri dish was unceremoniously pipetted in a salt solution onto a glass slide. For those who believe that human life begins with fertilization, Shettles' simple laboratory procedure was an act of unjustifiable killing, even though such experiments might help perfect a morally justified technique like genetic surgery. Even in the case of laboratory mistakes that might produce monsters, argues Bernard Häring, only those that are clearly inhuman should be destroyed. A number of scientists, on the other hand, subscribe to an alternate ethical view that an embryo is not human until later in its development—perhaps as early as two months or as late as six months.

Most scientists, naturally, fight what they see as arbitrary limits on their right to experiment. But not all. Testifying before the House subcommittee on science in January, Molecular Biologist James Watson took time off

"Don't laugh, Harkness—but every time I start an experiment these days, I wonder whether it's going to be the one where I end up finding religion."

from his cancer investigations to express concern about developments in embryo research. Predicting that many biologists would soon join Britain's R. G. Edwards in experimenting with human eggs, Watson suggested that one course of action could be to prohibit all research on human cell fusion and embryos. Failing that, he proposed international agreements limiting such research before it becomes widespread and irresponsible, and before "the cat is totally out of the bag."

Watson is not alone in his worries. Last summer Biologist James Shapiro, one of the three young scientists who successfully isolated a bacterial gene, gave up his promising career to take up social work because he feared government misuse of genetic achievements. An Episcopal priest, Canon Michael Hamilton of Washington (D.C.) Cathedral, called Shapiro's action a "loss of nerve." Yet the looming issues are enough to test the nerve of any thoughtful man. Central is the question: Who will decide? Who will make the choices not only of life and death, but what kind of life?

To consider such issues, Roman Catholic Lay Theologian Daniel Callahan and a number of like-minded ethicists and scientists have set up the Institute of Society, Ethics and the Life Sciences. Among the 70 members are Geneticist Theodosius Dobzhansky, Psychiatrist Willard Gaylin, Theologian John C. Bennett, and U.S. Senator Walter F. Mondale of Minnesota, who three years ago introduced a bill to establish an interdisciplinary committee to examine new scientific problems. It did not pass, but Mondale is trying again this year. "There may still be time," he says, "to establish some ground rules."

The long-term goal of the institute, says Callahan, is "legitimizing the problems," making the study of ethical issues a respectable part of the scientific curriculum. Too many scientists, says Gaylin, "see this as something mushy, something for Sunday morning, beyond the realm of science." To change that situation, the institute is trying to educate legislators on the importance of ethical considerations, and is encouraging universities to offer a solid background in ethical studies for "every scientific professional." At the Texas Medical Center in Houston, a similar interdisciplinary effort has been started by the Institute of Religion and Human Development and the Baylor College of Medicine. The Sunday School Board of the Southern Baptist Convention has developed a thorough adult-education course on biomedical issues as one of its electives for this spring.

Cancer Researcher Van Rensselaer Potter of the University of Wisconsin has suggested in a new book, *Bioethics,* that the U.S. create a fourth branch of Government, a Council for the Future, to consider scientific developments and recommend appropriate legislation.

Indeed, some form of super-agency may be the only solution to the formidable legal problems sure to arise. Already, laws relating to artificial insemination by a donor are in confusion; developments such as donor mothers and cloning will raise even more complicated questions. If a mother had herself cloned without her husband's permission, for only one example, would he be legally responsible for the child?

Some scientists, however, frankly believe that laymen are ill equipped to discuss issues with them, let alone share control of what they do. The matters, they contend, are technical and should be decided by the technical men who understand them. Even if government does enter the field, points out Daniel Callahan, much of the success of any ethical policy will depend on a responsible professional code. "If you depend solely on laws, sanctions and enforcements," says Callahan, "the game is over." Molecular Biologist Francis Crick is confident that basic morals and common sense will prevail. Some of the wilder genetic proposals will never be adopted, he claims, because "people will simply not stand for them."

Some ethicists and scientists argue that the worries, the plans and the proposals are premature, that ethics has always been an *ad hoc* thing, dealing with the world as it is, not as it might be in the future. Given the enormousness of the new problems and the speed of change, that attitude may be a luxury.

Beyond the sanctity of human life, the single criterion that ethicists most often mention as an absolute, or nearly one, is human freedom. Scientific advances, as they see it, can either promote freedom or inhibit it, but the distinctions are not always obvious or easy. The danger is that a democratic society might therefore fail to act at all, and by default pass the problems—and the solu-

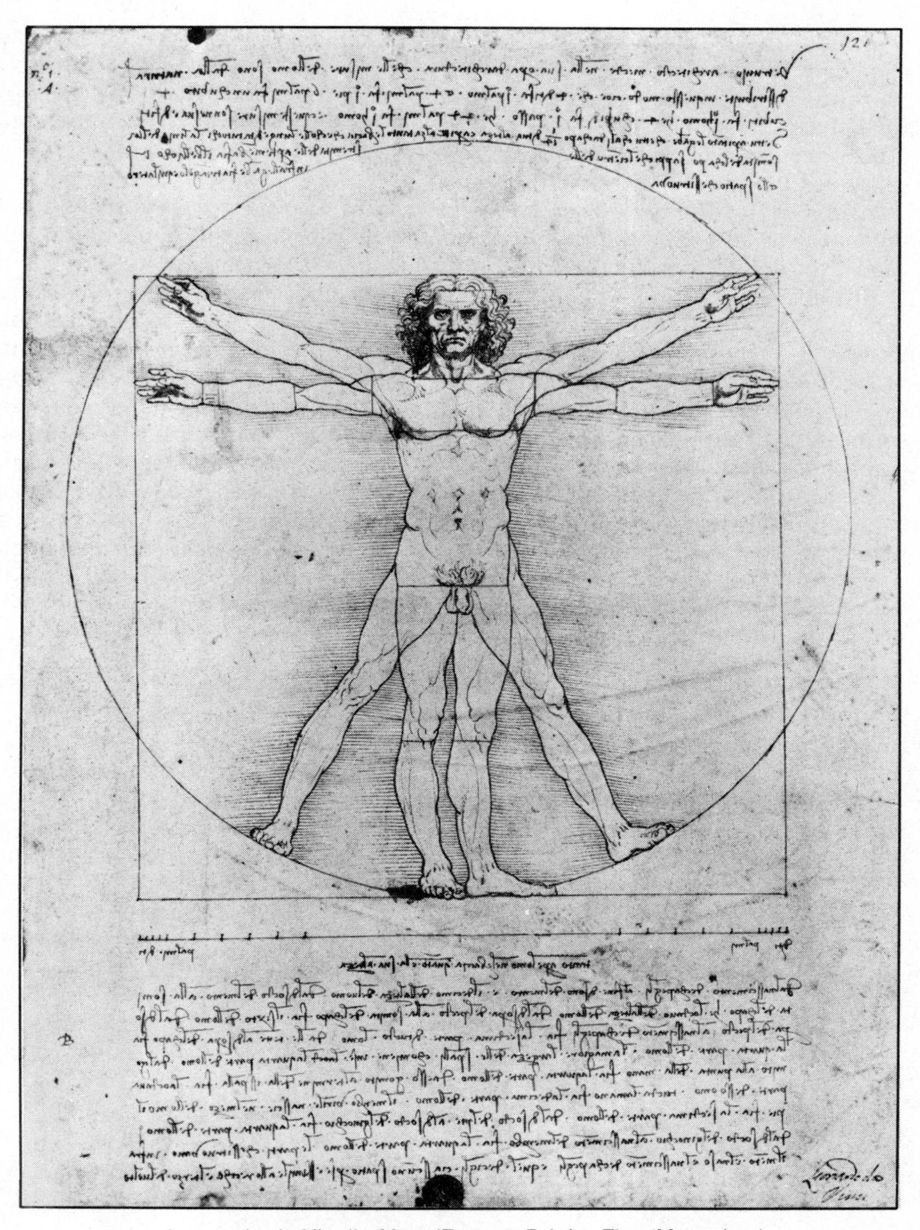

Leonardo da Vinci's *Man.* (Emmett Bright, *Time* Magazine.)

tions—to a small, uncontrolled elite, leading perhaps ultimately to a totalitarian government. The late author C.S. Lewis warned more than a quarter century ago that "man's power over Nature is really the power of some men over other men, with Nature as their instrument."

Despite the urgency, there can be no single ethical approach to the problems posed by the new genetics. The mechanists may want simply to deal with the facts of molecular biology, exploiting its discoveries as well as they know how, but not quite willing to look beyond to spiritual considerations.

Among many religious thinkers, there is an affection for the futurist philosophy of Teilhard de Chardin, who wrote glowingly of a coming scientific age when men would exult in "fathoming everything, trying everything, extending everything" on their road to an ultimate Omega Point of shared godhood. Finally, there are those, believers and unbelievers, who know man to be a victim of what might still be called original sin. Those in the religious community, especially Roman Catholics, warn that man must not tinker with such sacred values as life and the family for fear of disturbing the natural order of things. Those in the scientific world, more pragmatically, tend to mirror Potter's warning about "dangerous knowledge"—knowledge that accumulates faster than the wisdom to manage it.

There is hardly a chance for complete consensus among the three schools, but it may help to borrow a lesson or two from each. From the mechanist, his conviction that there is an order in the physical world, discoverable and manageable if it is approached with enough humility to comprehend its mysteries. From the Teilhardians, the confidence that God, whoever he is, has something to do with the future and may yet meet man there. From those who still believe in man's propensity for error, the willingness to put on the brakes a bit and reflect on values and consequences—but also, as Helmut Thielicke counsels, the courage to act despite almost certain knowledge that man will make serious mistakes.

As they look back toward the time when man stood on the threshold of a biological revolution, troubled and uncertain, but determined to push ahead, what will the beings of the future say about their ancestors? Caltech Biologist Robert Sinsheimer suggests an optimistic—and poignant—answer in his essay "The Mind of Pooh": "Perhaps, when we've mutated the genes and integrated the neurons and refined the biochemistry, our descendants will come to see us as we see Pooh: frail and slow in logic, weak in memory and pale in abstraction, but usually warmhearted, generally compassionate, and on occasion possessed of innate common sense and uncommon perception."

26
NEW HORIZONS FOR THE NEW DOCTORS

WILLIAM N. HUBBARD, JR.

. . . The marvels of modern science are known in detail by the new physicians with us and by intuition or common report by everyone here. Let me review briefly some of the major trends: Where descriptive theories have become exact sciences; much of psychology is being reinterpreted in neurophysiology; what was taxonomic in descriptive biology is now being restated with clearer insight in biochemical terms, while physiology and biophysics represent a trend to precision. We have been able to reduce complex living systems to components that are managable and that can be examined. The terms molecular biology, biochemical genetics and physiological psychology hint at some of these powerful instruments of examination. These are the most powerful analytic tools that have ever fallen into the hands of biological scientists in our whole history. They have been the most productive research strategies in terms of bringing understanding and new effectiveness to our efforts. They have, indeed, been the basis of objective, quantitative and reproducible descriptions which form a kind of reaction against the purely observational and anecdotal and inexact phases that science and medicine went through in their earlier formulation.

Out of all this has grown the optimistic assumption that a unified theory of material events may develop which includes all of the phenomena of living systems. One of the most influential assumptions of modern biological science is that the very best and indeed the only truly scientific approach to the study of a natural phenomenon is to divide the living system into fragments and to investigate elementary structures and elementary properties in greater detail. There is danger in this trend, though: that the reductionist analysis that the scientist must undertake will involve him intellectually and emotionally so deeply in the elementary fragments of the system, and in the analytical process itself, that he stands in danger of losing sight of the phenomenon of life which poses the question in the first instance. For the student

The Edwin Munich Memorial Lecture, presented to the second graduating class of the University of Kentucky College of Medicine, 10 May 1965.

of man and for the physician who ministers to man there is a sequence that we must guard against: one starts out with a question of great importance and relevance to human life, and then progresses *ad seriatim* to the organ or function involved, and then to the cell or cellular fragments, and then to molecular groupings, and then to individual molecules and atoms. And we would, I truly believe happily proceed – if we knew enough – to the elementary particle level where matter and energy are no longer distinguishable. Problems of great interest clearly evolve at each step of this reductionist approach.

But we stand in danger and must be forewarned against losing sight, as we proceed, of the important question that set us on this path in the first instance. It was Immanuel Kant who in a different context gave the warning that I would transmit to all of us today. Kant said that to yield to every whim of curiosity and to allow our passion for inquiry to be restrained by nothing but the limits of our ability . . . shows an admirable eagerness of mind, not unbecoming scholarship. But it is *wisdom* that has the merit of selecting from among the innumerable problems which present themselves those whose solution is important to mankind.

Science is supported because of its potential benefits for the human condition. The science of medicine, is, however, in potential conflict with the basic humanistic approach for which it is intended. There is no logical need to exclude the possibility of ultimate descriptions of all material events in chemical, physical, and mathematical terms. At the present time, however, this is not possible in states of highly organized complexity such as man. Our most convincing statements, scientifically, occur at an elemental level. Even when we move to the complex organic compounds our degree of certainty is decreased, and at the level of the cell and organ systems the certainty of our scientific description begins to waver. As we move to those important concepts of evolution and social systems, and indeed to the concept of a total ecosystem of the Universe, we find a failure in the prediction of events and descriptions of phenomena in physico-chemical terms which our modern science asks. These descriptions become less and less satisfactory as the *complexity* increases. The current notion then that physical concepts of causation, which are valid at the molecular level, will necessarily find themselves applicable at the most complex level yet remains to be demonstrated.

Next to complexity, one of the most important problems confronting science today is the issue of *uniqueness*. Modern science is really not primarily concerned with unique events, but rather with what they have in common with other events. Science deals with the unique only in terms of general principles. Literature, art, and history are generally concerned with unique experience. It is when science studies man that the tension between uniqueness (which is a characteristic of man) and commonality (which is a demand of science) becomes apparent. There are two ways of looking for truth. One is to find concepts that are beyond challenge because they are held either by faith or authority. And the second is to say that all truth as perceived by humans, even the highest, is accessible to tests; that by doubting our very understanding of the truth we are led to inquire, and by inquiry to refine our perception. Indeed, if science has a morality, it is the morality of self-examination. The foundation of this morality is a respect for truth and, therefore, it is bound to come into conflict with all attempts to curb freedom of thought and speech and, indeed, to introduce any form of authoritarianism. Moreover, scientific thought and its devotion to truth are themselves a product of an evolutionary process and, like all such products, must continually re-prove themselves in order to have survival value.

Complexity. Uniqueness. And the third problem that I would present to you is the one of *purpose*.

Scientific purpose is derived from the technology of its application. Many scientists, and indeed perhaps most scientists, are really not concerned with the application of the fruits of their immediate inquiry. There is nothing in the logic of the scientific method which can predict the manner of its utilization. Value systems, purposive work, artistic creativity, and the very idea of consciousness, fall largely outside of the main thrust of the modern life sciences and are unavailable at the present time for physico-chemical description or prediction. And yet within these purposive trends lies the distinction of man.

Increasingly during the recent decade the exact biological sciences have focused on the phenomena which are common to all the mammalian species and to all living forms. This trend away from the special purposive attributes, which particularize human beings, makes scientific biology appear far remote from human preoccupation. Nevertheless, the findings of orthodox biological sciences have profoundly influenced some of the largest philosophical expressions of modern humanism.

It is a paradox that the very success of comparative biology and evolutionary doctrine in relating man to the rest of creation may have retarded the growth of knowledge concerning man himself. Since all living forms have so many characteristics in common, biologists and even medical scientists naturally tend to focus their investigative efforts on organisms which are simpler than man and easier to manipulate in the laboratory. This tendency is based on the widespread—though unproved—assumption that the understanding of man will eventually emerge from detailed knowledge of the elementary structures and functions which are common to all living things. One of the possible deplorable consequences of this attitude is the common belief that the only fields of biology which deserve to be called fundamental are those that deal with the simplest manifestations of life and preferably with lifeless systems and structures derived from living things. Yet it is certain that such a limited approach is not sufficient to create a science of life, much less a science of man.

The evolution of the human race is now threatened. But it is threatened by a failure of social integration, not by biological evolution. This failure of social integration is as serious as if the individual members of the race had their nervous integrative systems becoming inadequate to the size and complexity of their bodies. The integration that is actually failing is a social function and represents a conflict between interests of national societies and the interest of common humanity. The conditions of this crisis have been partly produced by scientific knowledge and, therefore, by the activities of scientists. Because this form of knowledge is frequently misunderstood by nonscientists, the scientist now has a new and important social function.

This social function is emphasized when one appreciates that . . . the power of computers, together with modern means of communication, linked to knowledge of the psychology of persuasion—these may be used to create a tyranny over the mind of man more complete and unbreakable than any which has dominated human beings in the past. This capacity to influence man's mind on a scale and with an effectiveness that is unprecedented is not really inherent in the design of computers, but is rather a social phenomenon representing a selected use . . .

The essential problem is that we do not have at this time a comprehensive conception of the nature of man which allows us to approach him scientifically. The absence of such an hypothesis is a crippling disadvantage. I would suggest in closing that one of the essential ingredients of such an hypothesis would be the primacy of the private, personal, subjective, individual experience over any set of experiences of which science can now give a public and general account. This means that persons would have to be regarded as values in and of themselves and not as reducible to either physico-chemical systems or bundles of psychological trends or impulses. The social and political implications of this proposition are of some importance. One implication is that science, though an end in itself to the scientist, is only a means to an end where others are concerned; and that other end is to improve the possibility of a greater fulfillment of their identity as unique human beings.

27 FIGHTING EXISTENTIAL NAUSEA

DENNIS GABOR

We twentieth century liberals are assailed by many doubts. All around us in the Western industrialized countries we see a democratic prosperity without parallel. There are hardly any pawnshops left; the descendants of slum-bred proletarians flood the beaches in their millions. There have never been so many happy people about. But will it last? The happy people around us are mostly the first generation, who enjoy their prosperity by contrast with the poverty and insecurity they have known. There is also all around us the rising flood of crime, at the annual rate of 10 to 12 per cent in Britain and in the United States, and an even steeper rise in violence. All this and much more are before our eyes, and it is sufficient to make us doubt the simple faith of our nineteenth century predecessors.

But these doubts are not as new as one might think, nor were our nineteenth century spiritual ancestors as naïve as we might think. Here is one of the greatest of them, John Stuart Mill:

"Ask yourself whether you are happy and you cease to be so."

"If the reformers of society and government would succeed in their objects, and every person in the community were free and in a state of physical comfort, the pleasures of life, being no longer kept up by struggle and privation, would cease to be pleasures."

These are thoughts from Mill's autobiography, written a little more than a hundred years ago. The first two are from his period of doubt, which attacked him around the age of twenty. (Later he compared them with "the anxieties of the philosophers of Laputa who feared lest the sun should be burnt out.") But Mill, like Bellamy and Wells after him, had only to lift up his eyes and see the wretched proletariat around him to be fortified for his struggle. Many of our contemporary liberals do the same, only they lift their eyes a little higher and see the miserable poverty in Asia, Africa, and Central America. But pres-

From "Technology and Human Values," *Center* Magazine. Reprinted with permission from "Technology and Human Values", an *Occasional Paper;* a publication of the Center for the Study of Democratic Institutions, Santa Barbara, California.

byopia can be as bad as myopia. If we cannot offer any better hope to the developing countries than that in twenty or thirty years their gangsters will be able to make their escape in motor-cars instead of on foot, they will not be much encouraged on their way to becoming industrial democracies.

Existential nausea has always worried the rich; democracy has now put it within the reach of all. Some of the rich who were exceptionally gifted with will-power have directed it towards the public good; the English aristocracy, like the American money-aristocracy, has thrown up splendid leaders. For others, less favored, the escape from boredom often led to alcoholism and crime. This has now become only too general among our juvenile delinquents and semi-delinquents with their craving for "kicks." Among the older it finds a more innocent but rather sickening outlet in the bingo halls and at the slot machines, where they pull the handle just as a rat, whose pleasure center in its hypothalamus has been fitted with an electrode, works a handle for "kicks."

It is not surprising that many people prefer to shut their eyes to this aspect of the contemporary scene; only a few have dared to face it, like David Riesman who had the courage to write down the bold words: "What we dare not face is not total extinction, but total meaninglessness."

Life will always be full of meaning and interest for the creative minority and for healthy children. The technological society shows a sound instinct by fostering in grown-up people the mentality of children who want more and more toys. This of course is anathema to intellectuals, who have reacted with universal revulsion to Aldous Huxley's "brave new world," with its happy consumption-conditioned people. But to us, thirty-five years after Huxley's classic dystopia, with the Hitlerite and Stalinist horrors and the Second World War still fresh in our memory, and with the hydrogen bomb before our eyes, the prospect does not appear so terrible. The "brave new world" is here, and the question is, can it be permanent and stable? . . .

The problem before us is immense. It requires more knowledge of economics and of human psychology than can be fairly expected of any one human being. It requires also a "precognition" of the likely development of the sciences that have now become of vital importance: biochemistry and pharmacology. If I jump into the breach and present [a remedy which is] unavoidably amateurish, it is in the belief that [it has] a sound core, and in the hope that [it] will be gradually improved:

Non-permissive Education. I consider it as axiomatic that the age of plenty . . . [requires] higher ethical qualities than the ages of scarcity. Necessity forced people into the attitude of the honest, hard-working citizen; the police did the rest. Not much hardship was required in education (though the Victorians made it hard, for reasons of their own); the hard world put people on the straight and narrow path. When parents told their children, "If you do not work hard at school, you will starve, or become a common workman, etc." this was easily understood, though not always believed. Now they will have to tell them, "If you do not work hard at school and if you do not acquire the will-power to keep up the habit through life, you will be bored and unhappy." This will be neither understood nor believed, because children are seldom bored—except by work. The habit must be inculcated by conditioning at an early age. I do not believe in the modern tendency of educators, which is really more than a hundred years old, that the child should learn everything by playing, without an effort. Only exceptional people can learn without an effort. The vast majority are more likely to become televison addicts, or worse.

The technological society requires a trained intelligence, but it requires ethical qualities even more. Since the work of Alfred Binet and of Lewis Terman, we know fairly precisely what we mean by intelligence, but we hear only a rather vague conception of ethical qualities. An "ethical quotient," or EQ, is at least as much needed for the characterization of the social usefulness of a person as the IQ. In the Table [on the following page] I have tried to sketch out how an EQ could be constructed for *practical purposes* in the hope that somebody will complete it and put it into practice. . . .

There is of course a great difference between any such ethical scale and an intelligence scale. Ethical behavior is not problem-solving; it cannot be tested by any battery of questions. (The Greeks knew that it is no use asking a liar whether he is a liar.) Nevertheless, questions are not useless, if they are aimed at the views of the individual regarding *others,* not himself. Liars and cheats usually consider all people to be cheats and liars. If somebody thinks that he lives in the jungle, he is likely to be a jungle animal himself (though not necessarily a beast of prey). But reasonably reliable indications can be obtained only from observation of actual social behavior, and this creates another important difference between an IQ and the EQ. The first is a measure that can be applied to school children, the other properly only to adults, though the forecasts of experienced educators are often surprisingly accurate.

Note that this intends to be a *practical* scale. It does not matter whether exemplary social behavior is motivated by the active love of fellow-brethren, by unconscious inhibition, by vanity, by the fear of the police, by the fear of God, or by the love of God. An "assessment of merit" which takes these into account would be a matter for theologians rather than for social scientists.

Assuming now that something like this scale is accepted, what can we do with it? The most vital question for the sociologist is, of course, to correlate ethical behavior with environmental conditions (such as education, housing, etc.) and also with parameters that are not, or only little, influenced by environment (such as heredity or intelligence). In all cases the first thing is to put an individual in the right place on the scale. Here we can take the IQ as our model. . . .

The correlation of the EQ with the IQ is an obvious first application of the concept, and not without interest. As the IQ is something that is not easily altered, it goes some way toward deciding the old problem of to what extent crime is a result of nature or of nurture. What is more important, it gives the social planner a quantitative picture of the available human material. . . .

I am saying nothing new when I point to lifelong education, for men *and* women, as the way out. . . . The only positive contribution I can make to this is the emphasis on *non-permissive education* at an early age to steel the will-power, so that household chores, which some time will really diminish by labor-saving machinery and by improved public catering, will not leave men and women too exhausted for anything but watching television. They must be educated to become capable of prolonged development in their maturity and post-maturity.

This may also bring us closer to a much neglected ideal: *diversity.* Ours is a sadly uniform world, and it could hardly be otherwise in a democracy that is becoming prosperous. It will become even more so if it is crippled by overpopulation. But if we escape the triple danger of nuclear war, overpopulation, and existential nausea, born of inner emptiness, we may approach a worthwhile world, which will be rich in diversity and not just in consumption. We do not see the end, and we do not want to see it. Our slogan should not be Utopia, but the more modest maxim: *"The show must go on!"*

AN ETHICAL QUOTIENT

EQ Rating	Characteristics of Social Behavior
130 plus	Dedicated to good works and to the service of others to the point of self-effacement or even self-sacrifice.
120–130	Dedicated to socially useful works. Absolute refusal to act anti-socially, but ego not suppressed.
110–120	Socially unimpeachable behavior, balanced attitude between ego and social environment. Capable of unselfish behavior.
100–110	Responsible and reliable in the right environment, but prone to accept the standards of the majority.
90–100	Good citizen in routine conditions, but capable of mean, selfish acts. Occasional liar.
80–90	Social being under supervision, but capable of occasional dishonesty (not returning excess change, shoplifting, etc.), poor sense of ethical values, attracted to lower standards, fond of "kicks."
70–80	Inclination to envy, hatred, occasional cruelty, and criminal behavior. Prone to fall afoul of the law.
70 minus	Brutish, malicious, cruel. Habitual criminal.

28

TECHNOLOGY AND VALUES: NEW ETHICAL ISSUES RAISED BY TECHNOLOGICAL PROGRESS

HARVEY BROOKS

While a discussion of the origin and evolution of social and ethical values is beyond the scope of this paper, it is impossible to deal with the impact of technology on values without some assumptions as to the social function of values. In this regard I tend to adopt a rather pragmatic approach. I believe that the formation of value systems is an adaptation which enhances the survival of the social entities in which the individual claims membership. In this respect, values are the product of a cultural evolution, and result from natural selection through social and economic processes in much the way that the biological characteristics of species result from natural selection acting on variations in genetic constitution. What makes cultural evolution more complicated is that it is partly conscious and partly unconscious. Values are transmitted culturally, especially in the process of socialization of children, and this process is analogous to genetic inheritance. Different sets of values have different survival value both for the individual in his social milieu and for the social entity of which he is a part. Values change both because the physical environment of the society changes and because of social units in which individuals have partial membership change. The first of these processes of change has an analog in biological evolution, but the second is . . . unique to cultural evolution. A biological individual belongs to only one species, but a cultural individual belongs to many different social entities simultaneously, and this plurality of membership, plus the size and inclusiveness of the membership group, is characteristic of advanced societies.

For primitive man, the survival of the extended family, or, at most, of the tribe based on kinship, was the determinant of all values, and it was the relation of man to his natural surroundings and occasionally his conflicts with other similar groups which governed the values which would emerge with highest priority. Even at this stage, technology played a role in the evolution of values be-

From *Zygon,* Copyright 1973, University of Chicago Press. Paper presented at the Nineteenth Summer Conference of the Institute on Religion in an Age of Science, Star Island, New Hampshire, 29 July–5 August 1972.

cause of its influence on these relations of primitive man.

For modern man, technology plays much more of a role because of its capacity to alter the natural environment and because of the relationships of interdependence which it creates among the various systems of membership to which the individual belongs. According to Mesthene*, technology creates values both by creating previously unattainable options and by changing the relative costs of existing options. Social and group relations which seemed a part of the natural order a generation ago now seem within the capacity of man to change by the application of technology: for example, the elimination of poverty. Thus, a choice has been created which places the issue of poverty into the ethical domain.

The concept of the survival and welfare of the social entity is a subtle one, and becomes more complex as intersecting membership groups proliferate in modern industrial societies. The survival of the group may require the self-sacrifice of the individual, and many values relate to the choices which have to be made between the welfare of the more inclusive groups at the expense of the more restricted ones—the welfare of the nation versus that of the family or the welfare of all humanity versus the welfare of the nation-state. Such choices are frequently implicit rather than explicit, dictated by tradition and habit rather than by rational choice. Indeed this is essential, for our span of attention is too limited for us to afford to open up every issue anew every time it is presented to us as a choice. Similarly, the welfare or survival of the group is also an implicit rather than an explicitly articulated notion. As technology increases interdependence, the welfare of the group against its rivals becomes less important than the common welfare of the larger system which includes the rival groups, but it takes a long time for values to adjust to this more inclusive membership and achieve a new balance of choices.

The situation is further complicated by time horizons. To what extent should we choose to sacrifice comfort or well-being today for the very survival or well-being of our posterity? How is the future to be valued in relation to the present? The issue here is not simple to resolve, primarily because we cannot know enough about the future. As I will discuss in more detail below, technology puts us in a position to do many things that effectively commit future generations to courses of action or conditions of existence which they might not like, and which might even be catastrophic if we make some wrong guesses. This has also happened in the past, as for example in the irreversible ruin of agricultural land in the Mediterranean Basin, but it has never been possible on a worldwide basis before.

*Mesthene, E. G., *A Final Report,* Harvard University Program on Technology and Society, July 1972 (Cambridge, Mass.: Harvard Information Office, 1972), p. 166.

OVERLINKING OF VALUES TO TECHNOLOGY

Although it is obvious that technological progress has a profound effect on social values, there has been an exaggerated tendency recently to ascribe all such changes to technology. Because technological change has been rapid and highly visible during the past thirty years and changes in social values very rapid during the last decade, it has been tempting to assume a direct causal connection between the two. In an indirect sense this is probably valid. Increasing affluence and interdependence among industrialized nations would have been impossible without technological progress, and these factors have certainly been important in many of our value changes. On the other hand, I share the view of Mesthene that the actual consequences of technological change represent only a small subset of those which were theoretically possible. Technology is not a *deus ex machina* which develops according to its own inherent logic, independent of the values and preferences of the surrounding society. There is such a logic, and each new piece of technology is in a real sense genetically related to what has gone before. But cultural and economic selection strongly intervene to determine which small subset out of the many technical possibilities is actually developed and applied on a socially significant scale. Here again there is an analogy with biological evolution. The biological characteristics represented in each generation are inherent in those of the previous one, but what actually survives is governed by selection which has more to do with the environment. Similarly the totality of technical characteristics of each new generation of technology is determined by the technical logic of its genesis but only what is socially selected survives to govern the technical possibilities of the next generation of technology. Due to the imperfections of human motivation, comprehension, and wisdom, the technology that is socially selected has a large measure of randomness and accident, but it is still under human control.

As one surveys recent history, especially that of the last seven or eight years, one can make a fairly convincing case that our values have been changing much faster than our technology, and that, in fact, much of the current malaise is due to a lag of science and technology as well as institutions behind the expectations which have been generated by the social changes of the last twenty years or so. These expectations, in turn, were the result of the wholly different formative experiences of the present generation now coming to adulthood, as compared with my own generation. To the extent that these formative experiences were determined by television, suburban affluence, and extraordinary expansion of access to education, they may be said to stem from technological change. But the expectations themselves have far outrun the capacity of technology and sociopolitical management to deliver, at least on the time scale of the generational change.

One can, of course, argue endlessly as to whether the quality of life has deteriorated in the last twenty-five years. It is the fashion today to say there is nothing good about American society. But my own belief is that our perception of deterioration is due far more to the escalation of our expectations than to the deterioration of the objective situation of the majority of the American population. . . .

The central fact about modern technology is that its powers for both good and evil increase as it evolves, and thus place an ever greater burden on human responsibility and choice. By making possible the realization of previously abstractly stated and generalized ideals, technology confronts us for the first time with the full consequences of our goals, and with the conflicts and inconsistencies between them. Living with technology is like climbing a mountain along a knife-edge which narrows as it nears the summit. With each step we mount higher, but the precipices on either side are steeper and the valley floor farther below. As long as we can keep our footing, we approach our goal, but the risks of a misstep constantly mount. Furthermore, we cannot simply back up, or even cease to move forward. We are irrevocably committed to the peak.

29
DYING WITH DIGNITY

DAEL WOLFLE

Marcus Aurelius' assertion that "an emperor should die standing up" and the Western pioneer's wish to die with his boots on exemplify the desire to die with dignity. Increasingly we lose this opportunity. Progress in the prevention and cure of acute illness has shifted most deaths to the chronic disease category and has made lingering terminal illness more frequent. In earlier days, most people died at home or at work, tended by friends and family. Now the terminal patient has largely lost the security of dying in familiar surroundings, for most deaths occur in a hospital or nursing home, where medical skill and sophisticated equipment sometimes prolong vital signs after all hope of recovery and sometimes after sentience and self-control have disappeared. These capabilities are sometimes used, yet typically the treatment given the terminal patient is poorer in quality and quantity than that given the patient who is expected to recover, for the interest of the hospital staff is in saving lives and restoring health. No member of the staff has had professional training in dealing with dying patients, their relatives, or the problems of bereavement. All of this makes for added stress for the patient and his family. One study has found that in the year following the death of one member of a family, the death rate among close relatives is twice as high if the primary death occurred in a hospital or nursing home as it is if the primary death occurred at home. We have the curious situation that medical progress has made death more stressful for relatives, more expensive for the family, and more troublesome for society. Because these are discomforting matters, we have pushed them aside; death seems to have replaced sex as the socially taboo topic.

Yet physicians, psychiatrists, and sociologists are becoming more interested in the conditions and circumstances of dying. Among research findings is the demonstration of a significant dip in death rates just before patients' birthdays, before such major events as Presidential elections, and among

From *Science, 168,* No. 3938, 19 June 1970, p. 1403.

Jewish patients before the Day of Atonement. (Remember that John Adams and Thomas Jefferson both lived until the 50th anniversary of the signing of the Declaration of Independence, and died that afternoon.) This type of self-control of the time of dying poses few problems. More active controls—suicide and euthanasia—raise moral difficulties. And the physician's own increasing skill leads him into ethical dilemmas. When and for how long should he use heroic methods to continue life a little longer? Is a heart transplant worth the $20,000 or more it costs? Would a billion dollars a year be well spent on 50,000 heart transplants, with their frequently short survival times and high maintenance costs? Which patients get, and which should get, the use of scarce facilities that permit a few of them to live a few more days or weeks?

Physicians alone cannot answer such questions. They call for wider attention, for they all involve scientific, ethical, humanitarian, social, and sometimes religious considerations.

Is society ready to analyze death and the prolongation of life in terms of cost-benefit analysis, or to consider shifting the use of expensive facilities from the hopelessly ill to those whose future holds more promise? What about the customary reluctance to administer powerful but addictive drugs until "near the end"? What do we think of the "senseless prolongation" of life? Birth is no longer blindly accepted, but increasingly is planned and timed. Does this development and the growing acceptance of abortion indicate a readiness to consider euthanasia? The taboo against the discussion of such questions will have to relax, and seems already to be doing so. A society increasingly concerned about the quality of life cannot omit the final chapter from its concern.

PROGNOSIS

Rustum Roy and G. R. Barsch

People frequently tend to think of problems, or even of life itself, in terms of an *either/or* philosophy: that only a sharply drawn choice between two distinct alternatives is capable of solving or explaining all problems and mysteries. This was the uncompromising theme of Soren Kierkegaard's great theological treatise, "EITHER/OR," and there is little doubt that this philosophy has dominated much of modern western man's thinking. It is, in fact, intimately related to the development of modern science, for one of the foundations upon which the scientific method rests is Aristotelian logic, which embodies the classic *either/or* situation: of two conflicting statements, one and only one can be true.

And so today, living as we do in a scientific age, we find ourselves in everyday situations continually faced with what we perceive as sharp dichotomies. We may feel, for example, that *either* science *or* art must be the major outlet for expressing man's creativity, that true learning takes place *either* through the cognitive *or* the affective mode, that *either* heredity *or* environment determines behavior, that *either* our bodies *or* our minds determine our psychic health, or that *either* science *or* religion is the integrative force in human life, and so on. The literatures of education, of psychology, of sociology, and of philosophy are replete with impassioned arguments on both sides of these pairs of alternatives.

But we believe it is time to consider that many such dichotomies may in reality turn out to be totally false. Not only is it the proper time, but it is of the utmost urgency that western man start now to adopt a very different world-view: that he replace the *either/or* with a *both/and.* He must see *both* science *and* art as man's great creative adventures, recognize that *both* cognitive *and* affective modes are essential to learning, admit that *both* heredity *and* environment shape a person's ability to cope with the world, and perceive as complementary tools for gaining insight into the meaning of our world *both* religious tradition, with its vast store of empirical human wisdom expressed in symbolic language, *and* the powerful methods of science.

It appears to us as a sign of hope that in the important area of science and life, the false alternatives of *either/or* are slowly being replaced by a new synthesis, a new and fruitful interaction: the relationship between science and values. Until recently, it was the explicit aim of science to remain ethically neutral, to remain completely free of any and all external values. In the past, this ideal has served the necessary function of liberating science from the guardianship of political and religious authorities, thereby making possible the emergence of modern science. But since the coming of age of science and its rise to the dominant role it has achieved in the twentieth century, it has become mandatory that the further growth of science, through research and technology, be guided more and more by ethical principles. Thus we are witnessing today a wholly new dialogue between science and religion, a new convergence of science and values, of knowledge and responsibility.

But where are we to find the necessary guiding values for the further development of science? There are two main schools of thought: one that believes that the values reside inherently within science itself, and the other, which believes that the guiding principles are to be found in man's long history of inner and outer experience. Representatives of the first group are biologists J. Bronowski and J. Monod, and sociologist-psychologist R. H. Cattell.* An example of their position is the following quotation from J. Monod's *Chance and Necessity:*†

Modern societies accepted the treasures and the power that science laid in their laps. But they have not accepted—they have scarcely even heard—its profounder message: the defining of a new and unique source of truth, and the demand for a thorough revision of ethical premises, for a total break with the animist tradition, the definitive abandonment of the "old covenant," the necessity of forging a new one. Armed with all the powers, enjoying all the riches they owe to science, our societies are still trying to live by and to teach systems of values already blasted at the root by science itself.

No society before ours was ever rent by contradictions so agonizing. In both primitive and classical cultures the animist tradition saw knowledge and values stemming from the same source. For the first time in history a civilization is trying to shape itself while clinging desperately to the animist tradition to justify its values, and at the same time abandoning it as the source of knowledge, of *truth.* For their moral bases the "liberal" societies of the West still teach—or pay lip-service to—a disgusting farrago of Judeo-Christian religiosity, scientistic progressism, belief in the "natural" rights of man, and utilitarian pragmatism.

Where then shall we find the source of truth and the moral inspiration for a really *scientific* socialist humanism, if not in the sources of science itself, in the ethic upon which knowledge is founded, and which by free choice makes knowledge the supreme value—the measure and warrant for all other values? An ethic which bases normal responsibility upon the very freedom of that axiomatic choice. It prescribes institutions dedicated to the defense, the extension, the enrichment of the transcendent kingdom of ideas, of knowledge, and of creation—a kingdom which is within man, where progressively freed both from material constraints and from the deceitful servitudes of animism, he could at last live authentically, protected by institutions which, seeing in him the subject of the kingdom and at the same time its creator, could be designed to serve him in his unique and precious essence.

A utopia. Perhaps. But it is not an incoherent dream. It is an idea that owes its force to its logical coherence alone. It is the conclusion to which the search for authenticity necessarily leads. The ancient covenant is in pieces; man knows at last that he is alone in the universe's unfeeling immensity, out of which he emerged only by chance. His destiny is nowhere spelled out, nor is his duty. The kingdom above or the darkness below: it is for him to choose.

The second group includes scientists from very diverse backgrounds: paleontologist-priest P. Teilhard de Chardin, geneticist Th. Dobzhansky, anthropologist L. Eiseley, and biologist R. Dubos.* One of the most thoughtful statements of this position is the following quotation by the physicist-philosopher-statesman C. F. von Weizsäcker:‡

*Books written by these authors are included in the brief bibliography at the end of this section.

†From *Chance and Necessity* by Jacques Monod, translated by Austryn Wainhouse.

‡From C. F. von Weizsäcker: *The History of Nature*. University of Chicago Press, Chicago. 1949.

The conflict exists between two sets of facts both of which can be established objectively: on one hand, the instinct to fight our fellows, an instinct that is there for better or worse; on the other hand, the conditions that have to be met if mankind is to go on living. We can change our faith, but we cannot thereby escape this conflict. The conflict expresses the simple fact that man is a being who cannot remain such as he is. He cannot go back to the innocence of the animal; he must go forward to a new innocence, or perish. And the one possibility open to him, surely, is expressed in the words: God is Love.

What is this love? It is a transformation of man down even into his unconsciousness. It is a fusion of his insight with his instinct, a fusion that makes possible an attitude toward his fellows which was impossible before. The stirrings of man's instincts are the raw material. From this material, love builds a new person . . .

The scientific and technical world of modern man is the result of his daring enterprise, knowledge without love. Such knowledge is in itself neither good nor bad. Its worth depends on what power it serves. Its ideal has been to remain free of any power. Thus it has freed man step by step of all his bonds of instinct and tradition, but has not led him into the new bond of love . . .

But when knowledge without love becomes the hireling of the resistance against love, then it assumes the role which in the Christian mythical imagery is the role of the devil. The serpent in paradise urges on man knowledge without love. Anti-Christ is the power in history that leads loveless knowledge into the battle of destruction against love. But it is at the same time also the power that destroys itself in its triumph. The battle is still raging. We are in the midst of it, at a post not of our choosing where we must prove ourselves.

Irrespective of how much these two positions may differ on the question of the origin of values, they have in common the very important insight that the future of man—both of individual human life and of society at large—is an unknown and undeveloped territory that depends crucially on our decisions of today and tomorrow. The future is open. It can hold a great promise—the possibility of a new kingdom of meaning, fulfillment and riches—or an enormous threat—that one billion years of organic evolution culminating in the rise of man turn out to be abortive. It is for us to acquire the knowledge and understanding that are necessary to decide and to behave, not as passive objects, but as active and deliberate participants in the evolution of life and love.

REFERENCES FOR FURTHER READING

Bronowski, J.: *Science and Human Values*. Harper and Row, New York, 1965.
Cattell, R. H.: *Beyondism: A New Morality from Science*. Pergamon Press, New York, 1973.
Dobzhansky, T.: *The Biology of Ultimate Concern*. New American Library, New York, 1967.
Dubos, R.: *A God Within*. Scribner's Sons, New York, 1972.
Eisely, L.: *The Firmament of Time*. Atheneum, New York, 1962.
Monod, J.: *Chance and Necessity*. Alfred A. Knopf, New York, 1971.
Teilhard de Chardin, P.: *The Phenomenon of Man*. Harper and Row, New York (1959). Teilhard de Chardin, P.: *Science and Christ*. Harper and Row, New York, 1968.
von Weizsäcker, C. F.: *The History of Nature*. University of Chicago Press, Chicago, 1949. von Weizsäcker, C. F.: *The Relevance of Science*. Collins, London, 1964.

Photo courtesy of Dr. Robert L.

SECTION FIVE

SCIENCE, GOVERNMENT AND SOCIETY

The Problem

Is the scientist a hero or a villain in society?

We all agree that the applications of scientific inquiry to practical affairs have made tremendous changes in the lives of everyone. There is hardly a place on the globe, for example, where you will not find and hear a transistor radio. And this little radio is based upon devices that were totally unknown only 25 years ago. Developments in information theory, electronic circuit theory and switching devices have made possible the ubiquitous high-speed computer. And improved understanding of the nature of chemicals provides new drugs, better fertilizer production methods, and the imminent possibility of modifying the genetic makeup of animals and people.

But the *application* of science, guided by scientists, destroyed Hiroshima and Nagasaki along with more than 100,000 human beings in 1945. And today, automobiles of the latest design are hardly 25 per cent efficient in the utilization of gasoline's energy, while they pollute the atmosphere with hydrocarbons, lead and nitrogen oxides so that the sun grows dim and our eyes grow red. The federal Food and Drug Administration, it appears, has to police drug firms vigorously and constantly to keep untested, unneeded, and even unsafe medical chemicals from being widely distributed. And this still is true in spite of the fact that pharmaceutical firms maintain well-appointed laboratories of their own and spend a substantial sum each year on scientific research.

There are those who would join the Nobel Prize-winning chemist E. B. Chain when he says, "The activities of science are morally and socially value-free. Science is the pursuit of natural laws, laws which are valid irrespective of the nation, race, politics, religion or class position of their discoverer. Although science proceeds by a series of approximations to a never-attained objective truth, the laws and facts of science have an immutable quality. The velocity of light is the same whoever makes the experiment which measures it. Because this is the case, although the uses to which society may put science may be good or evil, the scientist carries no special responsibility for those uses, save as a normal citizen. The two-edged sword of science is fashioned for whomsoever will pick it up and wield it." (Professor Chain speaks out of his experiences as a scientist, which included major contributions to the development of penicillin.)

On the other hand, Francis Bacon in the seventeenth century saw science as a source of power. Even today, nations encourage and support their scientists and laboratories in the interests of national power and prestige. The exploration of the moon has been made a national priority as much for national prestige as for the advancement of knowledge. Of the nearly $20 billion that the United States Government puts into research and development each year, more of it goes to develop military weapons of destruction—albeit with the claim that they are the guarantor of peace—than into any attempts to reduce human infirmities or to deepen our knowledge of nature.

This section begins with some intriguing comments on the nature of modern science, as perceived by Kenneth Boulding, Professor of Economics in the Institute of Behavioral Science at the University of Colorado. His diverse activities range from research on the achievment of peace to the writing of sensitive poetry about the human condition. The selection chosen for inclusion here develops the notion that science is a subculture of our society, and that the two unique characteristics of the scientific subculture are the high values it places on veracity and curiosity. Although these traits may not be valued as highly in other parts of our society, they do contribute substantially to the revolutionary effects of science. In Dr. Boulding's words, "The scientific revolution and science-based technology represent a kind of take-off from the old world of classical civilization."

In the second article of this Section, John Ziman examines the question of whether scientists should be responsible to society for the consequences of their discoveries. John Ziman is a professor of physics at the University of Bristol in England, where he is much concerned about communicating the significance of science to citizens. In the essay included here, he argues that for him a significant social duty of scientists is to act as informed and independent critics who can protect society from technical disasters that may arise through error or through the cupidity of powerful men or institutions. To insure that scientists can play this role, Ziman says, it will be necessary for universities to keep governments and industries at arm's length—an attitude that may well stretch the ability of university administrators beyond what they are prepared to do.

A briefly-presented example of a situation that illustrates John Ziman's thesis is contained in Article 32, a news item prepared by the Washington correspondent of the British scientific journal *Nature*. It concerns an order by the U.S. Government for bakers to improve the nutrition of Americans—apparently whether they want it or not—by raising the iron concentration in all bread. At the time this news item was prepared (February 1974), hearings had been scheduled for the purpose of considering certain related safety questions. "The hearings will thus be a scientific debate over the likely effects of increased iron intake. . . ." Participating in the debate, which might still be in progress as you read this, will undoubtedly be a number of university scientists—one of our society's habitual major sources of expert opinion. As you will see, they do not always agree.

Might vigorous, large-scale scientific research activity carry with it the seeds of its own destruction? In Article 33 Alan Mussett, a geophysicist at the University of Liverpool, presents us with a startling argument: that the current rate of scientific discovery cannot long continue without exhausting all knowledge, and that science must soon therefore find itself with nothing left to discover. If this is a correct assessment, what will happen as the scientific well begins to run dry? Will our society stagnate as a major source of its innovative drive becomes exhausted? Mussett's thesis, it should be pointed out, is not likely to be acceptable to most scientists, since they have already seen the end of science predicted frequently and confidently over the past hundred years.

Whether or not there may be an end to science at some far future time, John Platt urges scientists in Article 34 to solve some of the urgent problems confronting society *now*—lest humanity itself come to a tragic and sudden end. Dr. Platt is a distinguished biophysicist in the Mental Health Research Institute of the University of Michigan. His aim is to set forth the major problems which, if not solved, will cause large-scale degradation of life or even total annihilation of the human race. Once we recognize what the problems are, the work of scientists, engineers, managers and politicians can be intelligently organized. Dr. Platt says, "The task is clear. The task is huge. The time is horribly short. In the past, we have had science for intellectual pleasure, and science for the control of nature. We have had science for war. But today, the whole human experiment may hang on the question of how fast we now press the development of science for survival." Professor Platt's plea, in contrast to Mussett's views, gives little room for worry about science's running out of problems to tackle.

Of course, no one denies that scientists are social beings and that science is a social activity, but the question of whether science serves society at large or only the private ends of a few is one which can well be argued. Lee DuBridge was for many years President of the California Institute of Technology; more recently he has been science advisor to the President of the United States. He qualifies, as do few people, as a true elder statesman of science. He believes science to be a servant of society: "Whether we look around the world or look backward in time, the vast benefits of scientific knowledge are clearly evident." Article 35 by Dr. DuBridge is a practical and authoritative assessment of science's role in our society, written for his fellow American scientists.

As you read these essays, you will be weighing in your mind the benefits of science against the problems it creates. And perhaps, as you read, you will be measuring the validity of certain simple world-views, as represented by astrology, demonology, contemplative meditation and other reputed remedies for the world's ills, against the growing complexity of our minds and bodies and of the world itself, as revealed to us more and more by modern science. Do you find a body of verifiable experience and ideas that illuminates your own personal quest?

Laurence E. Strong

30 THE SCIENTIFIC REVELATION

KENNETH E. BOULDING

There is a certain implicit assumption today that science is something above and beyond society, a kind of genie out of a bottle, which promises or threatens to do all sorts of good and bad things to us, but which belongs, as it were, to another order of creation. But this view of science as a genie outside of society, whether angelic or demonic, will not stand up to serious examination. Even though the rise of science might have something of the impact of a "revelation" in sociological terms—that is, as a creation of evolutionary potential which is realized as the years go by—it is still a revelation which is very firmly embedded in human society and must be visualized as a phenomenon taking place, as far as we know, wholly within human society.

We have to regard science as . . . an expanding movement within . . . the social system. The subculture of science began with a small group of people in Europe in the second half of the sixteenth century, suffering some persecutions—on the whole fairly mild—and having to exist first in something of an underworld. One can perhaps date the chartering of the Royal Society in 1662 in London as the first great legitimizing act, with Charles II as the Constantine of science. From this point on, there is no doubt about the legitimacy and respectability of the scientific subculture, even though from time to time it comes into conflict with other subcultures in the society, such as the church and occasionally perhaps even the state.

"CHURCH OF SCIENCE"

From its small beginning, science, like other great [movements] has expanded until it is now worldwide in scope and enormously influential. Scientists, indeed, now constitute a "clergy" about as numerous as the religious clergy and certainly better paid and much more powerful. In the 1960 census in the United States, there were 197,000 clergymen. Male chemists, natural scientists and social scientists amounted together to

Drawing by Vladimir Rencin

179,000, and women add another 29,000, just topping the clergy at 208,000 total.

In spite of its successes, science remains a fairly small subculture. The people who think of themselves as scientists, who read scientific journals, who try to keep up with their own field, who teach or do research in some area, do not number much more than one in a thousand of the American population and of course a very much smaller proportion of the world population. Like the clergy, scientists have something of a congregation of laity—the students that they teach, the engineers, doctors, social workers and other professionals who look to one or the other of the pure sciences for the theoretical base of their technology. Even the laity of the scientific "church," however, probably do not amount to more than one per cent. It is all the more understandable, therefore, that the majority of the population regards science as something wholly outside them, an alien force, even in a sense as an alien religion which they often perceive indeed as something of a threat to their own folk culture.

In the conflicts of the scientific subculture with other subcultures around it, such as the conflict with the churches over evolution and the conflict with the Communist Party in the Soviet Union over genetics, the academic community tends to assume that the scientific subculture has always won hands down. This may be in part an illusion fostered by the relative isolation of the academic community from the rest of society. In the United States, for instance, church membership has risen from about seven per cent of the population at the time of the American Revolution to 64 per cent today. Membership in the "church of science" is harder to define, but two per cent would be a wildly optimistic figure . . . Even in the socialist countries, where science is officially elevated above religion, it can suffer serious political persecution when it seems to run counter to official ideology.

ANTI-SCIENCE

We should not even underestimate the potentialities for substantial anti-scientific popular movements, even in the West. It is true that there is a kind of peaceful co-existence between science and religion in most countries of the world today, a co-existence based mainly on sheer segregation and the absence of communication on both sides. The hostility towards science, however, among fundamentalists, both religious and political, is a strong undercurrent which could easily break through to the surface. Furthermore, we now see a secular anti-scientific movement among the young, especially among the Hippies and the New Left, which easily slips over into astrology and other forms of superstition. The popularity of astrology in the United States, indeed, is quite a testimony to the isolation of the scientific subculture and the very superficial impact which it has on the majority of people.

THE SCIENTIFIC ETHIC

Every subculture has an ethic, or at least a set of common values and preferences. The scientific subculture has had a highly characteristic ethic which has been remarkably persistent and on the whole remarkably well observed. Its origins are obscure and puzzling, as it is actually a rare ethical system, although apparently quite stable once it is established in a subculture. Perhaps its most striking characteristic is the high value which it puts on veracity—that is, abstaining from deliberate lies. The one sin against the Holy Ghost in the scientific community is the publication of deliberately falsified results. Cultures which put a high value on veracity, however, are quite rare. One has to look, perhaps, for puritanism, whether in its Protestant or in its Catholic form, as the source of the ethic of veracity which has been so important in science, but this still remains a very puzzling feature of intellectual history.

Another high value in science is curiosity, although it is not highly regarded in many folk cultures—as folk proverbs indicate, curiosity killed the cat. In political cultures, especially in the international system, neither veracity nor curiosity are highly regarded. A diplomat, indeed, is one sent abroad to lie for his country and an incurious loyalty is regarded as much preferable to the asking of embarrassing questions. In the military subcultures, also, veracity and curiosity are very little regarded and, conversely, the military virtues are of very little use in the laboratory. The religious subcultures, too, foster value systems which in many cases are at variance with those of science. There may be something a little monastic about the traditional devotion of scientists to their work. Indeed it can be argued that the peculiar tradition of Christian monasticism with its emphasis on the sacredness of work *("Laborare est orare")* and the insistence of Christianity, by contrast with Eastern religions, on the reality and sacredness of the material world created a climate out of which science could develop. Scientists, however, are not particularly noted for chastity or poverty, and they have a positive distaste for obedience. The bourgeois ethic is perhaps closest to the scientific ethic among all the surrounding subcultures. The insistence on calculation, accounting, the careful use of time, the pragmatic attitude towards life (if it doesn't sell, don't buy it) and especially the puritan bourgeois ethic, which despised bargaining and chicanery and set fixed prices, were all favorable to the development of the scientific ethic.

CONFLICT

A conflict between science and the rest of society still arises in part because of the conflicting ethical systems in the different subcultures. There is increasing unhappiness in the scientific community with secrecy, with the sort of deceptions which international politics seem to demand, and with the military ethic. It could well be that in the next generation we shall see a conflict between science and the military state as severe and as acute as the conflict it had in earlier centuries with the church, especially where the demands of the state for its own survival go counter to the interests even of its own citizens and, still more, the interests of the world as a whole. The military state then may become an enemy of its own citizens, and doubly the enemy of the scientific community.

Perhaps the most difficult ethical problem of the scientific community arises not so much from conflict with other subcultures as from its own success. Nothing fails like success because we don't learn from it. We learn only from failure. One could argue indeed that the very success of the scientific community is a result of the fact that it succeeded in legitimating failure and hence removed the main obstacle to the growth of human knowledge, which is the refusal to learn from failure because of the threat which this poses to our identity as a person. In the scientific community the propositions and theories of the scientist were divorced, at least in part, from his status as a person. We see this in one of the great moral myths of the scientific community: the scientist is pleased when his particular theories are disproved. The fact that in practice his pleasure may not be entirely unalloyed does not diminish the importance of the principle.

This very toleration and legitimation of failure has produced a stupendous success, the witness to which is the great science-based technology which has developed since the middle of the nineteenth century, the great symbols of which are the decline in "infant" mortality (under the age of 70), nuclear weapons and the voyage to the moon. This enormous success has given man the power either to destroy himself or to move forward into a quite different state of human life, which I call the "developed society," this being presumably what we get as a result of the process of development. . . .

SPACESHIP EARTH

The one thing we know about a developed society is that it has to inhabit a "Spaceship Earth." It is well recognized that our existing technology is fundamentally suicidal, resting as it does on a linear process which begins with the extraction of exhaustable resources in the shape of ores and fossil fuels and ends in pollution. The great unsolved problem of technology is that of creating what is being called a "looped" economy in which man finds a comfortable life in the middle of the process which is essentially circular, that is, in which the waste products of human activity are all used as raw materials for the next cycle of production. We are still a very long way from this kind of technology, although there are the beginnings of it in, for instance, the Haber process for the fixation of nitrogen from the air (1913) and the Dow process for the extraction of magnesium from the sea. Ultimately, it is clear we will have to use the atmosphere, the oceans and the soil as inexhaustible material resources in the sense that what we take from them we will also put back into them. . . .

A very important question . . . is that of the mutual interaction between the scientific subculture and other subcultures of society, especially the political and the folk cultures. There is a certain tendency within the scientific community to assume that all that is necessary is an expansion of the scientific subculture into more and more areas of life. This view is at best a gross oversimplification and at worst a dangerous illusion. In the first place, the scientific subculture, and the technological "super-culture" which it has produced, is not and probably cannot be a complete culture. It is true that there is a world superculture of, say, chemistry practiced with much the same symbols and ideas by chemists everywhere. No matter what the ideology of the surrounding society, chemists will all have the same mandala in the shape of the Periodic Table on the walls of their classroom and will be proclaiming much the same universal truth. As soon as the chemist steps out of his classroom and laboratory, however, he becomes an American or a Russian, a Catholic or a Protestant, a Maoist or a Hindu, an Africaner or a Kikuyu. He is rarely a chemist for more than 8 to 10 hours a day. The rest of the time he is immersed in his domestic and his local culture, of which he may be a slightly aberrant member, but from which he will probably not diverge too sharply.

We may doubt whether the scientific subculture has penetrated any society as deeply as Christianity penetrated medieval Europe, or as Islam penetrated the culture which it created, though this admittedly would be hard to prove. It seems true, however, that those countries which have been most successful in accepting the scientific superculture, and in generating the kind of economic development which is based on it, are also societies which have had a strong and vigorous folk culture, as in Europe, the United States and Japan. Where the folk culture produces an ethic which is ill-adapted to the modern world, as it seems to be in the Arab states, the very impact of that superculture disorganizes a society rather than moving it toward development. What we have to think of, therefore, is much more of a symbiosis between the scientific subculture and the other subcultures with which it is surrounded and with which it interacts, rather than any sort of conquest of the other cultures by a kind of universal church or culture of science.

The problem of how to create this symbiosis is a very proper problem for social science and one on which, as far as I know, very little work has been done. The critical problem here is that of the impact of the various subcultures on each other, particularly in regard to their value systems. The scientific subculture and related technology have produced an enormous impact on all other subcultures—whether it is the family, the church or the state, the military or the arts, or the youth, the middle aged or the aged—simply because human values have a very slim genetic base and are mostly learned. . . .

In recent years there seems to have been a small tendency for scientists, or at least a small group within the scientific subculture, to become more selfconscious about the mutual relationships between the scientific subculture and others. . . . On the whole, however, the ivory tower tradition of science is still very strong, and those scientists who are concerned about the impact of the scientific subculture on others are still regarded as a little odd.

Perhaps the next generation will change all this. One of the most encouraging signs of the times is the extraordinary mobilization of youth in questioning the established values of virtually all subcultures of all societies. While this questioning can degenerate into nihilism or a retreat into superstition, it can also force us into painful reappraisals of many of the things that we have hitherto taken for granted. It questions the subservience of the scientific community either to the state or to commercial interests. It insists that the only ultimate product of technology that makes any sense is the good person and the good life, however this may be interpreted. It questions anything that seems to be exploitative or cruel. It rediscovers the virtues of tender-mindedness in human relations, which is certainly not inconsistent with a tough-minded attitude toward the truth.

We recognize grave dangers in this movement. It could lead to monstrous perversions, as the youth movement in Germany was perverted by Hitler. If the scientific community, however, is sensitive to the fact that it is not the only subculture on the beach, and that it must maintain subtle inputs and outputs and even bargaining relationships with the other subcultures around it, there is a good chance that this increased awareness of the world may enable us to avoid the traps with which the whole developmental process is increasingly beset.

The scientific revolution and science-based technology represent a kind of take-off from the old world of classical civilization. The "flight" of development cannot go on forever. At some point there must be a re-entry into Spaceship Earth. This re-entry will present acute difficulties. If, however, we have a clear view of the nature of the problem, a certain optimism about our power to solve it is entirely reasonable. The one great cause for optimism indeed is the clear fact that the evolutionary potential of the human nervous system is very far from having been exhausted and that there is no [reason not to expect] continued human learning. Human learning is the key to all our social problems. . . . The possibility that we might find out something about human learning which would enable us to accelerate it is an even greater reason for longrun optimism. . . .

31
THE IMPACT OF SOCIAL RESPONSIBILITY ON SCIENCE

JOHN ZIMAN

It is not given to us mortals to perceive the full consequences of our actions. Moral responsibility is therefore an issue that cannot be decided by scientific procedures. When an evil is traced back to a cause that we have freely created, we can almost always produce an excuse.

THE ESCAPES FROM RESPONSIBILITY

'I had no idea that this would happen, Sir!'—thus, perhaps, Rutherford, for splitting the atom. 'It was all in a good cause, Sir!'—the discoverers of DDT. 'If I hadn't done it, Sir, somebody else would!'—nuclear fission, shall we say. 'They were going to do it, so I thought I had better do it first!'—the biological weapon makers. '*They* made me do it, Sir!'—a general excuse for all servants of all corporate bodies. 'We tried it out, and it *seemed* to work all right!'—the thalidomide tragedy. 'I didn't actually do anything myself; we just talked about it and the other chaps went and did it'—a compendium of justifications for all academic research. 'Well, it does make rather a mess, doesn't it, but everybody wanted to play with our new toy'—which covers much of the pollution problem. And so on.

Every such excuse is valid, however essentially infantile. The scientist is not in the front line, pulling triggers and dumping defoliants. By definition, he is a Back Room Boy, employed to discover principles and to design devices, not to hurt other people with them, so you can't really blame him for what has happened. True enough; but if he had not made that misused discovery, or if he had imagined its consequences, or if he had not allowed himself to be employed by that evil corporation, then perhaps the tragedy would not have taken place.

In the complex of social institutions within which we try to make ourselves at home on earth, the mind and professional expertise of the individual scientist are not a negligible force. The enormous size of the technical

Reprinted from *Impact of Science on Society, 21*(2): 113, by permission of Unesco, community seems to guarantee anonymity and to countenance irresponsibility, yet the intellectual leader carries ten thousand of his colleagues with him in a 'break-through', and sets ten million humble labourers on a new course of manufacture, commerce and use. We scientists cannot take personal pride in the 'achievements' of our science and technology and simultaneously repudiate responsibility for its failures and abuses. We are either humble workers in the vineyard—or we are indeed the New Men come to make a better world.

The dilemmas of personal responsibility are not new, and the history of ethics tells plainly that they cannot be resolved. Think of the Inquisition—or of the revolutionary turned executioner—before you put your trust in an ideology or a pledge of virtue or a Hippocratic oath. Gospels, social blueprints and other formulae acquire their legalistic interpreters, until the call for peace becomes a war-cry and the stake is an instrument of mercy.

I see no salvation in resolves or resolutions, however aptly phrased and apparently benevolent. To accept them without reservation is essentially to abjure responsibility; it is the abandonment of judgement and a flight from rationality itself. . . .

THE MAKING OF SOCIALLY RESPONSIBLE SCIENTISTS

Social responsibility in science rests therefore upon the way in which scientists are made. Whether or not we have the inborn talents for success in research, we are moulded by upbringing and education. But social responsibility is not a subject to be learnt from a course of lectures (Tue. 10, Fri. 10, Room G.44; Dr. Piravetz: Practicals, Sun. 2-5, Trafalgar Square). It is not something one can practise ostentatiously, as an example to the young ('I think I shall go out and do some social responsibility in science this afternoon. Anyone coming with me? We could count it as part of your optional field work'). It is an attitude of mind, a sensibility of the spirit implicit in an educational system, in personal relations, in institutional policies.

What is missing from the education of present generations of scientists? First, they lack general education. They go out into their corporate laboratories as learned ignoramuses knowing all about nuclear magnetic resonance, or the physiological function of adenosine triphosphate, but without any grasp of history, of philosophy, of political thought or of economics—or of other fields of science. The microbiologists scorn ecology, the nuclear physicists know nothing of warfare, the mechanical engineers are totally ignorant of the physiology of respiration, and so on.

In the mad rush to produce completely trained specialists on the cheap, we assume that they will somehow pick up the rest of the knowledge they need. How? From newspapers, from bar-room gossip, from television programmes? . . .

The student protest movement can certainly not settle, by brickbats and obscene invective, the dilemma of the county council in choosing between a new hospital, a better bus service, or an improved sewage works. The problems of social responsibility always arise in highly technical contexts, where expert opinion is pitted against expert opinion in the language of cost effectiveness, budget deficits, manpower projections, ton-miles per litre, perceived decibles and other jargon.

But we must somehow sensitize these earnest owlish experts early to think of people, of pain, of freedom, and of beauty. When they are middle-aged and grey-haired, and with the power at last to make such decisions, it will be too late. Sophistry and calculation will then have taken over: the spiritual lobes in their ponderous noddles will have died for sheer lack of exercise.

A CERTAIN HYPOCRISY

. . . Of all the despicable traits of the modern academic, nothing is worse than the hypocrisy with which he will receive money for research from organizations whose prime ends he inwardly detests. The biochemist who works on a project paid for by a military agency, knowing that it is somehow connected with biological warfare but pretending to himself and others that it is all good, clean, pure science, has sold himself to the Devil: it is really a pleasure to see such characters now getting hurt.

I don't mean that scientific work for military ends is itself immoral. Pacificism is an admirable doctrine for life and can only be respected, but so also can the principles and practice of self-defence against crime and violence. In civilized countries we support our armed forces, though we may deplore their need and regret their expense. To be a soldier may not be everybody's choice, in peace or even in war, but it is not a dishonourable profession.

What must surely be said, however, is that the scientist who has contracted to do research on behalf of a ministry of defence has become a technical soldier. If his country goes to war, then he, too, is pulling the trigger and dropping the bomb. He may, as a good patriot, feel that he is doing right: that is a question worth quiet debate in each instance. But he cannot take the money and still claim the privileges of a civilian. In civil wars, people get shot for much less than that.

Hypocrisy and opportunism are learnt at the mother's knee. The best we can do is to shame those who practise them.

CONFLICTS OF LOYALTY

We must learn, in fact, to clarify the essential conflicts of loyalty in the practice of science. For the old-fashioned academic it was easy: his loyalty was given first to his subject, then, in principle, to humanity—although a good rousing war-cry would quickly awaken his patriotism. The only really important thing to do in life was to add another few papers to the literature of 'microteleonomy' or 'macroscopology', which was itself, of course, of profound cultural significance because it was 'there'. It was not quite such a noble calling as it was sometimes represented, but once one had got oneself established one certainly knew who, what, and where one was.

Nowadays we have to think of our employers—who need to make profits or war, of our country—which ought to make peace, of our

"On the other hand, my responsibility to society makes me want to stop right here"

From Sidney Harris, American Scientist.

professional association—which ought to make up its mind, of our students—who need to make good, and even of our families—who ought to be made to shut up. I don't pretend to know my way through these thickets of obligations; one just tries to balance them up as best one can. But part of our moral education in responsibility must be an attempt to analyse these loyalties, lining them up in order of priority. . . .

SCIENTIFIC 'ANTIBODIES'

But let us not despair. The historical answer to the tyranny and irresponsibility of individuals and institutions is not merely personal martyrdom, cunning accommodation, or the preaching of unheeded sermons. The wisdom of society creates the countervailing corporate power: Parliament to curb the King, the courts to curb banditry, trade unions to curb the exploitation of labour, and so on. The 'balance of power' model, the adversary principle, the peculiar institution of 'Her Majesty's Loyal Opposition', are examples of a technique that could well be copied. If we have bodies of scientists dedicated to essentially irresponsible and selfish ends, of profit or power, then we must invent 'antibodies' to neutralize them.

This is the theory of the Food and Drugs Administration and other regulative agencies of the United States Government. Despite many short-comings, they do, in fact, perform their allotted functions. In principle, if imperfectly in practice, they oppose the collective powers of teams of scientists to the teams in the manufacturing corporations, thus acting as deliberately conscientious elements on behalf of society.

Within the framework of scientific methodology, this is the correct procedure. The task of the scientific innovator is to persuade the other members of the scholarly community that he has made a valid contribution. The task of the others is to oppose, not out of mere conservatism and prejudice, but as informed critics unwilling to be convinced by mere assertions. The progress of knowledge is dependent upon such debates—not elevated into personal controversies, and always tempered by tolerance and scepticism on both sides. The apparently certain tone of each particular scientific paper belies the underlying uncertainty of the issues; it is the duty of each participant to use his persuasive powers to the full, just as the duty of the advocate is to make the best of his case before the court.

Scientific responsibility in social issues therefore demands such debate, whether in the comparative privacy of the learned journals or in the form of the press or Parliament. In most cases, there is no absolute truth to be determined: DDT is both a blessing and a scourge; motor vehicles are convenient, but noisy and dirty. A balanced report by a single expert commission, however well intentioned, cannot judge between conflicting opinions and priorities until each party has expressed, to the utmost, its own special viewpoint or interest.

What is important is that there should be adequate representation, of a skilful, professional kind, of the interests of the general public, of conservationists, of bird-watchers, of the lovers of peace and quiet, of preventive medicine, of anglers, of the League against Cruel Sports, of the preservers of churches and windmills, besides the immediate economic contestants, such as industrial corporations, local authorities, public utilities, transport undertakings and so on. It is as much the duty of the State to ensure that this type of expert evidence and advocacy is available as it is in a criminal trial to have proper lawyers for the defence.

THE UNIVERSITY: SEAT OF RESPONSIBLE DISSENT

But where will such experts come from? Who will normally employ them? Industrial corporations and the various organs of the State know how to provide themselves with scientific consultants; what about the numerous other interested parties, especially the ordinary citizen who is to be assaulted, battered, deafened, poisoned, driven from his home, or just insulted? In theory, he, too, is the ward of a benevolent State, that will develop further agencies for consumer protection, planning, noise abatement, etc.

But the conservatism that breeds in such agencies is all too familiar. They continue to plod solemnly back and forth on ancient sentinel duties, and are blind to new dangers. Being in the political domain they are the target of political forces, such as those of corporate industry.

And who will guard the guardians? Where are the professional experts to countervail the military power, the powers of State monopoly, of national health services, etc? The American debate about anti-ballistic missiles (ABM) is a case in point. Although this is not at all a matter of pure science, the public case against the ABM system had to come from a small volunteer group of academic scientsts (who are not, indeed, ignorant on the subject) rather than from an engineering organization charged specifically with these interests.

Where can institutionalized, licensed dissent and criticism survive, except in our universities? The great issues of academic freedom and independence are more desperate than ever, now that the learned men have real fire to play with. Having demolished the ivory tower in the name of social relevance, we must rebuild it as a watch-tower over all matters of technique and social action. We need ABM-watchers, and CBW*-watchers, and pollution-watchers, armed not with slogans but with searchlights and telescopes of specialist knowledge.

This is not an easy exercise in social engineering. Your natural ABM-watcher is a cinch for a fat contract for secret work on radar systems for the Pentagon; a really good pollution-watcher is the ideal consultant on the payroll of an oil company; nobody knows anything about CBW except those who have already been sworn in under the Official Secrets Act. Yet I think it must be attempted.

The advantage of the university environment is that it provides a safe professional base for expert criticism. The academic is paid primarily to be a teacher and a scholar, not to provide specific research results for specific corporate bodies. A tenured professor does not have to be personally brave to say what he likes about the government, the local big wigs, and the policies of Generalized Manufacturers Unlimited. True, he may not get his next research contract renewed, but why should he care: it will give him more time to write a book instead of filing innumerable reports and questionnaires.

The academic economists, in their splendid running battles with Treasury policy, set us a good example. Chaps like John Kenneth Galbraith and Nicholas Kaldor move in and out of government circles—and when they are out of favour they don't hesitate to cock devastating snooks at their lords and masters in Washington and Whitehall, respectively. How many aeronautical engineers or biological soldiers are doing just that? This is not a matter of unusual individual 'responsibility'; it is just the institutionalization of loyal opposition in the academic watch-towers.

*Chemical-biological warfare (Ed.)

The real problem is how to switch the so-called research interests of many university teachers into such channels, and how to fund them so that they have adequate assistance and equipment without coming under the sway of precisely the corporations and agencies they are meant to watch. This is a field for much more experiment by bodies like the Ford Foundation and the National Science Foundation in the United States and their equivalents in other nations.

A SOCIETY GETS THE SCIENTISTS IT DESERVES

In the end, a country gets the scientists it deserves. A responsible society breeds, trains and fosters responsible scientists. An open market in ideas and political criticism is also open for technical attack and counter-attack. A free press, for example, ready to publish informed articles on scientific matters, may be the essential atmosphere in which kites may be flown as first signals of a storm of controversy. Remember that the single most important act of scientific responsibility in our time was the publication of Rachel Carson's book, *The Silent Spring*—surely impossible in a totalitarian society where all scientific questions are automatically solved by appropriately planned groups under the benevolent eyes of the all-wise Party and an omnicompetent Council of Ministers.

The problems of technological progress are not, in the end, capable of decision by 'scientific' methods. They are problems of social priority, of aesthetic judgement, of taste, of preference, of material and spiritual standards. The final arbiters must be the common people, as users and abusers of our pretty toys . . . .

SCIENTIFIC RESPONSIBILITY FOR TOMORROW

It may be, indeed, that the pendulum of prejudice has swung too far. The public adoration of the scientist, as the sage and saviour, is a thing of the past; now we seem to hear nothing but scorn for his pretensions and hatred of his arrogance. The movement to harness every technical expert to environmental studies or systems engineering, to make him useful and safe—your friendly neighbourhood boffin—could be as damaging as the older snobbery of pure science for its own sake.

Scientific knowledge and social action are not the same thing. The neo-Marxist argument that all science is 'really' determined by social ends is either a vacuous truism or it is dangerous nonsense. Natural philosophy is not entirely for useful ends, despite the technological spin-off. If you try, too shortsightedly, to press it into service for immediate ends, then you will rob later generations of its products. Countercyclically, I feel the need to preserve the collective skills, the expert knowledge, and the delicate social organization of the scientific community from the pressures of an ignorant public, a shameless press, rapacious money-makers and opportunist politicians. A certain aloofness, a slight distance from everyday affairs may be the only way of preserving these islands of sanity in a crazy world, not as refuges but as watch-towers and safeguards against far greater evils.

That is the paradox: social responsibility in science must not be too concerned about today, for tomorrow also will come.

32 FDA HALTS SCHEME TO COMBAT ANAEMIA

COLIN NORMAN

A controversial scheme to provide what amounts to preventive medicine through supermarkets and grocery stores has been called off at the last moment by the US Food and Drug Administration (FDA). The scheme involves adding large helpings of iron to most of the bread baked and sold in the United States, in an effort to combat iron deficiency anaemia.

Under pressure from influential medical and nutrition organisations, the FDA issued a regulation in October [of 1973] which would have tripled the amount of iron in so-called 'enriched' bread and flour. The regulation, which has the force of law, was due to be brought into effect on April 15 [1974], and [by February,] most bakeries [had] already made plans to implement it. But a barrage of complaints from individual physicians, concerned about possible danger to people suffering from rare blood disorders, led the FDA to announce . . . that it will turn the whole matter over to a public hearing. . . . The regulation will not be enforced until the safety question has been cleared up, the FDA said.

The idea of adding iron to bread and flour has influential support from the American Medical Association, the Food and Nutrition Board of the National Academy of Sciences, the White House Conference on Nutrition, and virtually every national nutrition association. With such an impressive pedigree, and a goal with which few would disagree, why has the scheme met with so much opposition?

The dispute is, on the surface, simply a straightforward disagreement among scientists and physicians over the interpretation of facts and the results of various clinical and epidemiological studies. But it also touches on some fundamental principles underlying the FDA's approach to regulating the food industry, and it calls into question the federal government's role in trying to improve the nutritional standard of the American diet. In any case, it is an interesting case study of regulatory decision-making.

The idea of fortifying basic foods with iron

From *Nature,* (London) *247:* 498, 1974. Reprinted with permission.

compounds gained impetus in the 1960s from studies which indicated that anaemia is widespread in the United States, particularly among women and children from low income groups. Since bread is the most widely consumed processed food, it became the obvious candidate for a fortification programme designed to increase iron intake, and in 1969 the idea was floated by the White House Conference on Nutrition as one of its many recommendations for improving the health of the American public.

A year after the White House Conference published its report, the idea was converted into a formal proposal from the American Bakers Association and the Millers National Federation, the baking industry's trade associations. They asked that the legally required amounts of iron in so-called enriched flour and bakery products be increased significantly from their present levels.

The FDA handled the proposal with some caution, first by turning it over to the American Medical Association for advice, and then by coming up with a proposal of its own. It is the FDA's proposal which was issued in final form [in] October [1973,] and which will be the subject of public hearings. . . .

In short, the proposal would require that, in order to bear the label 'enriched', all flour sold in the United States must contain 40 mg of iron per pound, and enriched bread must similarly contain 25 mg of iron per pound. According to the FDA, some two thirds of all bakery products now consumed in the United States are enriched. And, in view of the rapid progress of the fortification bandwaggon during the past few years—a phenomenon which has turned most breakfast foods into little more than high-calorie vitamin pills—there is a strong commercial incentive for bakeries to produce goods which can bear the 'enriched' label.

If the proposal is adopted and made a legal requirement, it will not be the first time that basic foods have been loaded with extra nutrients. For many years, vitamin D has been added to milk and iodine has been added to salt, and even iron has, for the past 30 years, been baked into bread. But the proposal represents a radically new policy. The present requirement is that enriched bread should contain between 8 mg and 12.5 mg of iron per pound, simply to ensure that the iron lost during milling of white flour is replaced; it is a policy which is followed in most industrialised countries. The proposed levels, however, represent a deliberate attempt to increase the average American's daily intake of iron.

According to the FDA, even at the proposed levels, the amount of extra iron that would be included in the average diet would amount to less than 20% of the recommended daily requirement. Thus, the agency argued when it published its final regulation, the scheme can hardly be called 'supermarket medication'. The proposal, the FDA said, strikes a balance between providing as much of the recommended daily allowance as possible, and ensuring that the risks to people suffering from blood disorders are kept to a minimum.

But critics of the scheme are not so sanguine. First, they have criticised the studies which led to the theory that anaemia is prevalent in the United States, and they have argued that there is need for more accurate information on the epidemiology of iron deficiency anaemia before such a fortification scheme is begun. Second, they have cited a sheaf of studies which cast doubt on the extent to which iron incorporated in bread is assimilated by man; again, they suggest that more information is needed before iron salts are loaded into bread. And they have also pointed out that nobody has yet managed to quantify the risks of increased iron intake to people suffering from such blood disorders as haemochromatosis, Cooley's anaemia and liver disease associated with alcholism. In other words, even though the idea of fortifying bread with iron was first proposed more than five years ago, the chief argument being put up against the scheme is that it is premature.

As for the prevalence of anaemia, the FDA bases much of its case on a study carried out in the 1960s, called the Ten State Nutrition Survey. This study indicated widespread anaemia, particularly among people from lower income groups, and it also came up with the finding that the problem is not solely confined to women and children. The methods employed in the study have, however, been criticized by several haematologists, who have argued that the conclusions may be false, and also that, even if anaemia is as prevalent as the study indicates, inadequate iron is not the sole cause.

In response to such doubts, the FDA set in train a reappraisal of the literature, from which its experts concluded that "there is a strikingly high incidence of iron deficiency anaemia in many segments of the US population." They also concluded that the studies have "consistently shown higher anaemia prevalence rates among blacks compared to whites, in low income states compared to high income states, and in low socioeconomic groups compared with groups higher in this regard." And those conclusions have been endorsed by virtually all the major medical associations in the United States.

But even if the studies are correct, would an iron fortification scheme help? Again, the FDA and the professional medical associations say that it would, and some individual physicians say that such a conclusion is not necessarily supported by the evidence. Writing in *Nutrition Today,* for example, Dr. Maxwell Wintrobe, Professor of Medicine at the University of Utah, and a highly respected haematologist, points out that a number of studies have indicated that iron salts added to bread are not readily absorbed by the body. In particular, a study carried out in Wales by Dr. Peter Elwood, which involved two large communities, produced "no conclusive evidence in terms of an effect on circulating haemoglobin levels", Wintrobe asserts. Wintrobe also pointed out that the signs of anaemia picked up by the ten state survey may be put down to a variety of causes, and thus simply increasing dietary iron intake can only solve part of the problem.

The FDA, however, does not concede that the iron fortification scheme would be either unnecessary or ineffective. Dr. Alexander Schmidt, Commissioner of the Food and Drug Administration, specifically stated . . . when he announced that the proposal would be suspended, that "several tests, including measurements of iron serum, iron absorption, and iron binding capacity have been utilized to verify that low haematocrit and haemoglobin values observed in nutritional studies establishing the prevalence of anaemia . . . are due to iron deficiency." Schmidt also cited two studies to show that man does utilize iron furnished by fortified bread. Thus, the FDA has announced that it will not consider further objections on those two points—the only recourse opponents have if the regulation is imposed is to go to court—and the hearings . . . will thus be concerned solely with the narrow question of risk.

The hearing* will thus be a scientific debate over the likely effects of increased iron intake on people suffering from storage diseases, but two other considerations are likely to arise. The first is that additional dietary iron could mask the diagnosis of diseases which are usually signalled by the onset of anaemia. Cancer of the colon, which causes chronic internal bleeding, is the most frequently cited example. And second, it has recently been suggested that intake of additional quantities of iron could lead to the development of Parkinson's disease in some people, because the disease is associated with increased iron stores in the brain.

The hearings will thus be concerned solely with the risk side of the risk-benefit equation. That is, however, unlikely to satisfy many of the critics of the scheme, who are asking for a full inquiry into the philosophy behind the proposal, and the need for the federal government to require by law that iron be added to bread.

As Dr. Wintrobe suggests, if poor nutrition is indeed the cause of iron deficiency anaemia in the United States, a more effective policy should aim at attacking the causes of poor nutrition. Such corrective measures as improving the economic status of lower income groups, reducing unemployment and lowering food prices would be more effective than "to cover our eyes, shut our minds, triple the dietary iron and hope for the best", he says.

---

*The hearings were held in June, 1974. As this book goes to press, the matter is in the hands of the Commissioner for final decision. (Ed.)

33
DISCOVERY: A DECLINING ASSET?

ALAN MUSSETT

We hear many warnings these days that soon we shall run out of oil, mineral resources, water, or even air. To those who can spare time to worry about other matters (and these are good times for worries) I offer a more spiritual concern: that in a few generations we shall run out of ideas to discover—perhaps chief of these being the discoveries of science.

Many will object that I am not the first to claim that science has nearly run its course, that scientists of the last century thought the same, only to be confounded by the discovery of the electron, X-rays, radioactivity and so on. What reason have I to suppose I am any more correct?

There is a simple test we can apply:—if science were complete, then it would follow that we could explain the perceptible world about us. Let us try it and see. "I look through the window into a garden where the plants are becoming green with a new spring: young shoots are bending up to the light, a blackbird is singing tunefully from the top of a pole, and there is even the golden gleam of an early crocus. Children are playing in the sun: some are squabbling over who are to be 'fathers and mothers' and who 'children', while others are running round with arms outstretched, pretending to be aeroplanes. But I don't think they will be playing much longer, for heavy clouds are piling up, and already I hear the rumble of distant thunder."

That's enough to begin with, as Humpty Dumpty remarked to Alice. How is it I can see through the glass but not the walls, how do the plants know when it's spring, how do they know which way to grow towards the light, why does the blackbird sing, what is the function of a flower and why are there so many kinds, how does the sun warm us, why do children play and what is the purpose of their games, why cannot we fly like birds, what causes thunder and lightning? This is only a selection of possible questions and I omit the problems of why we should find flowers and bird-song beautiful, though it seems quite possible to me that an under-

From *New Scientist 60:* 886, 1973. This article first appeared in New Scientist, London, the weekly review of science and technology.

standing of aesthetic appreciation will be found.

First, let the Victorians attempt each question. "Glass? Ah, yes, refractive index 1.510, made essentially of the elements silicon and oxygen, a super-cooled liquid, you know." But this is little more than a description and a recipe; and giving the refractive index sidesteps the issue by assuming transparence. It doesn't tell us why most solids are opaque. Nowadays we would explain in terms of energy levels and absorption bands. We can even go some considerable way towards explaining how we see: as well as the simple optics of the eye we know something of the chemical changes produced when light is absorbed by the retina; we know that the retinal cells don't send simple signals to the brain via the optic nerve but instead are so connected as to recognise lines and other patterns in different orientations; we even know that vision partly develops after birth in response to the type of things seen.

Next question: thunder and lightning. "Oh, that's easy enough to explain. Lightning is a discharge of electricity at high potential, and the sudden release of energy produces the flash, while the consequent rapid expansion of the air is propagated at the speed of sound to your ears as thunder. There, sir!" Not at all a bad answer, and a great improvement on primitive ideas that it was the wrath of the gods or clouds banging together. But modern knowledge can elucidate the complex stages of charge build-up during the growth of a thunder cloud, and also the way in which the discharge propagates. Our knowledge of the weather has even reached the point where people admit that the forecasters are no longer wrong all the time.

"The blackbird: of the order passeriformes, of the phylum chordata. Warm-blooded, egglaying, nests in trees; eats insects or fruit . . .". Again, largely description, though the system of classification depends on genuine similarities and differences. The Victorian scientist would, of course, have known about evolution and the advantages for survival of warmbloodedness, caring for young and so on, but would have known little of animal behaviour, and would not have appreciated that the blackbird sings—not to delight us—but to "stake out" a territory. We know this is very common in animals though the advantages of territories are not fully understood.

NOT PROPER QUESTIONS

About the other questions, I believe the scientists of the last century could have said little, and I suspect that they would have felt many of them were not "proper" questions at all. In contrast we can say a lot. Plants recognise the time of year by the length of day, a fact we have exploited to produce chrysanthemum flowers all the year round. Shoots bend towards the light because on the sunny side growth is inhibited by a hormone induced by sunlight. Flowers have to compete as much as animals, and the flower aids reproduction by encouraging fertilisation via a symbiotic relationship with bees that depends upon a bee visiting only flowers of one species. We know not merely that the Sun maintains its heat by thermonuclear reactions, but have so refined the calculations even for the enormous temperatures and pressures inside the Sun that we are upset by discrepancies in the neutrino flux. Children play to develop their bodies and minds, to learn by copying their elders and to improve their cooperation. And although we cannot fly because our power-weight ratio is too low, we have overcome this lack with aeroplanes.

This, of course, is a sketchy answer to a small selection of questions that could be asked, but I hope it will remind you that we know a great deal about the world around us. We know much about the properties of matter even in the solid state and have exploited our knowledge in carbon fibres, transistors and so on. The chemists can analyse and often synthesise molecules of enormous complexity, including organic molecules vital to life, such as insulin or vitamin B_{12}. Medicine has banished many diseases and

even understands some of the disorders that arise from the misproduction of minute quantities of subtle substances in remote organs of the body; physiologists understand the body in great detail and can even measure the electrical response of a single neurone; the genetic code is understood in principle and is being busily decoded; the locations of genes are being mapped, often by hybrids of human and mouse cells; people are currently studying how a fertilised egg develops into all the specialised organs of the body; studies in animal behaviour and anthropology give illuminating glimpses of our own social habits and behaviour; we have even got so near the inmost nature of creatures that we can modify behaviour by implanted electrodes or chemicals: for instance, we can turn "killer rats" into docile ones and vice versa by giving them chemicals; we appreciate the processes that over the vast periods of geological time shaped and continue to shape the surface of our globe; and the mists that shroud the origins of the solar system and life are no longer quite opaque.

There are blanks, certainly, becoming greater as one progresses from the natural sciences towards those sciences most intimately connected with man (which is not surprising for many of the discoveries related above required the electron microscopes and so forth of the physicist and the techniques of the chemist); moreover, it is easier to be objective about the electrical resistance of a new alloy than about the psychology of a person. We do not yet properly understand cancer, for instance, nor can we regulate the economy nor make people happy.

All in all, I think we of today can claim high marks in my test, whereas, if the Victorians had thought of applying it they would have been more struck by their ignorance than their knowledge. The test is about the world around us and not the carefully selected problems of the laboratory, and I think we score highly, not just in comparison with 100 years ago, but absolutely: that there are few questions that would stump us completely.

A further objection that might be raised is that there may be completely new areas of knowledge, not merely undiscovered but not even suspected—as, for instance, radioactivity was unsuspected 100 years ago. Of course it is impossible to rule out completely such discoveries but I think it unlikely for two reasons.

First, the test already proposed, if carried out systematically, will show up a number of gaps in our knowledge, but will also roughly define the limits of our ignorance. It might prove, of course, that in investigating these areas major surprises will occur, and the bigger the blanks the greater the chance of this happening; but the present framework appears adequate at present.

Secondly, for an area of knowledge to remain unsuspected it must impinge but minimally on our everyday world or else its existence would have been suspected already; but if it is to be discovered at all there has to be some situation—some combination of, perhaps, extreme conditions—in which it will show up as a discrepancy from accepted laws of nature. A simple test that can be applied is to ask if there have been any branches of knowledge that have been discovered by chance and that otherwise would probably have remained unknown. Areas in my own subject of physics suggest themselves as the sort of thing I am discussing: radioactivity, relativity, particle physics. But before I consider these as examples, it is necessary to consider what we mean by "understanding" and this leads to the concept of a hierarchy of scientific knowledge.

HOW SCIENCE BEGINS

A science starts by collecting data, i. e., describing exactly what is observed, and a law such as Boyle's Law is not much more than a convenient way of summing up how the volume of a gas is observed to vary with pressure, i. e., it is an empirical observation that does not seek to relate the behaviour of gases to other phenomena. Later the concept of atoms and molecules, together with certain assumptions about their properties, allowed the gas laws to be deduced through kinetic theory; also the chemical laws of constant composition, simple and multiple proportion became explicable. At this stage we have "explained" or related together a number of empirical relationships in terms of an underlying concept—in this case, atoms—which we consider to be at a more fundamental level. (We have begun to answer "why?" as well as "what?") As we add more and more to this fundamental level there comes a point at which we try to explain it by a yet more fundamental level below it; and so on down, level by level. At each step of the process the systematics of one level point to the existence of a lower level.

As an example, the systematics of the behaviour of [matter] pointed the way to a discrete atom. In turn, the systematic behaviour of atoms . . . pointed to an atomic structure, though of course it didn't suggest what the structure is, nor how it could be discovered. But as soon as the atom was shown to consist of a positively charged nucleus plus electrons it made nuclear structure probable and so pointed to particle physics. (At present, particle physicists are looking for component particles of the so-called elementary particles.)

The point of this line of reasoning is that the discovery of a yet more fundamental level of understanding is not a new area of knowledge in the sense required (intellectually exciting though it is) but is more the result of the steady systematisation of knowledge which is one definition of the object of science.

It follows that particle physics cannot be regarded as a branch of knowledge that could only have been stumbled on by the chance discovery of radioactivity or some similar serendipitous event. . . . In a similar way we can regard the theory of relativity as an example of a new level of understanding required to systematise a number of diverse and inexplicable facts. An example that probably would meet the requirements is extrasensory perception and similar phenomena, but it has yet to be proved that these exist.

It may seem that I am dismissing all scientific advances as inevitable, and I suppose I am. It seems to me that the field of scientific knowledge is fairly well mapped out, and though chance will play a big part in how and when particular areas of knowledge are discovered I don't think it will affect the outcome. (It is generally conceded that no individual scientist is indispensable.) It may be that we shall come up against some apparently inexplicable fact, such as that living bodies have a "breath of life", and which we shall be forced to incorporate in our knowledge, but I see no need of it yet.

The next step is to assess how long it will take to exhaust science. We are told that the output of knowledge—or at least of scientific publications—is at present doubling every 15 years, and also that 90 per cent of all scientists who have ever lived are alive today. This exponential increase cannot go on indefinitely, and in fact seems to be lessening now; but even if there is no increase and effort remains at the same level it is still going very fast. If knowledge is doubling every 15 years, and just suppose we have reached the halfway point, then it follows that a quarter of all knowledge was discovered in the last 15 years. Thirty more years at the same rate will see it finished. It is not possible to know how far we have travelled towards completion, but I guess a quarter to a half, leaving 30-90 years to go.

This assumes a linear advance of knowledge with effort, which is unlikely, and I think some areas of knowledge will be discovered at an accelerated rate, while others will show decreasing returns for effort. An analogy will help here: let us imagine the totality of knowledge [as] . . . a little like a bookshelf in a library. Related laws will be placed close together so that one part of the expanse will comprise physics, an adjacent one chemistry and so on. Any areas of unsuspected knowledge, such as already discussed, would appear as almost detached lobes. The total is, of course, finite, an assumption not only fundamental to my argument but to science itself, for it has to assume that the natural world can be understood. (It is the understanding that is finite, the data can be infinite: the . . . variation of the volume of a gas with pressure, for instance, provides an infinite amount of data, but the law—that it varies inversely—is finite.)

MAPPING THE SURFACE

Going back in time to a period when little was known, this "knowledge surface" would be an almost total blank, and scientists of that past time could start almost anywhere: classifying flowers, dissecting the human body, observing the effects of heat on solids, and so on. They would each map a little part of the surface, and the boundaries of these little areas would define the limits of knowledge of each subject. . . . [These] early scientists would find that each problem solved would uncover several more problems, a well-known experience of scientists. But this would not continue indefinitely, for there would come a point at which much of the surface was mapped and the "explored" areas would join together: the situation would then be similar to the late stages of solution of a jig-saw puzzle when pieces suddenly drop into place at an accelerated pace. I believe that there are signs that this is happening. Many hybrid sciences have appeared in recent years, such as geophysics, biochemistry, and neuropsychology, in all of which several traditionally different lines of investigation come together and are seen to be aspects of the same thing.

Not all branches of science may benefit from this effect, for areas such as particle physics probably cannot be approached from more than one direction, and could prove steadily harder to elucidate if vaster and vaster accelerators are required.

Other types of scientific knowledge also will face diminishing returns. These are the sciences that deal with the inaccessible in space or time, of which I shall take geology as an example. A geological map can be regarded as a statement of what rocks and structures are in an area at present, and therefore in principle could be mapped to completeness, elucidating finer and finer detail with more and more effort, which would itself be an example of diminishing returns. But like any other science, geology seeks to understand the underlying causes that have changed the Earth, and in this sense the geology can also be regarded as a record of the history of the Earth, and as the record is not complete there will be parts of that history that can never be more than partially elucidated. In this sense the (historical) facts are not fully discoverable.

A very important area of discovery is invention, the creation of devices and substances that do not occur in nature. This has been a very fertile field, and it is through the inventions of technology—the wheel, the steam-engine, telecommunications, plastics, etc.—that science has had much of its impact. Some inventions do not require a very advanced supporting technology, for instance the wheel and ships, but most of the innovations of today do seem to require it: telecommunications required vacuum technology and now requires the expertise and understanding of semi-conductors; aeroplanes require sophisticated design and sophisticated materials, and so it is with other inventions. If this is true for the majority, the end of scientific discovery would result in a steady drying up of invention as the possibilities are exhausted.

Two areas may be exceptions. The first is computers: many of the features of computers were thought out by Babbage in the last century, but he was unable to realise a computer because it had to be mechanical, and so again it would seem that an advanced technology is required. This is certainly true in part, and computer improvements depend to a considerable extent on the reliability, cheapness, compactness and low power demands of solid-state circuitry. However, the fastest growing area of the computer world is not the computers themselves but software,* which already is the bigger business in terms of money. Thus the future of the use of computers may be partly independent of technology, and lie closest to developments in logic and concepts, traditionally the province of mathematics and philosophy rather than engineering. Perhaps computers could be intimately linked to human brains, increasing our capacities in ways more fundamental than the ability to calculate at prodigious speeds.

The second area is living matter, particularly intelligent matter. Once the first simple cell existed, one might regard all subsequent life forms as developments, in the sense . . . that one can trace the development of wooden ships from outrigger to galleon and tea-clipper, all of which used sails and a wooden hull. The great variety of life forms and subjective possibilities that have opened up for man—the enormous breadth

*A term used by computer people to indicate all the associated programs, procedures and documentation; as opposed to *hardware:* the computer equipment itself. (Ed.)

of experience that embraces curiosity, moral and aesthetic senses, love, friendship, the nobility of self-sacrifice and so on—has appeared so late in evolution that one wonders what might be achieved by a higher organism. But this fascinating topic is outside the scope of this article, and is one that would involve a creature that would not be man as we know him.

If my thesis is correct, does it matter? On this I can only speculate briefly (partly because of the missing half or three-quarters of scientific knowledge), but a possible danger that has prompted this article is that our society and culture may stagnate. We live in an era of rapid innovation that has been going on long enough to lull us into thinking it is the normal state of affairs. Yet history tells us that this state is not a typical one. There are strong forces of conservatism in any society, and at present they are not very evident because of the incessant barrage of innovation that forces change on us. It was not so long ago that "new-fangled" was as much a term of opprobrium as now "new" is used as a selling line for nearly everything except whisky. Even so, the past centuries in this country were far from stagnant; it is only that they changed at a slower pace than today's. These centuries saw the mental revolution that demoted man from his cosy position at the centre of the universe and then later from his special position in creation, leaving him comfortless in the void.

STAGNATING SOCIETIES

However, history gives us examples of civilisations or parts of society that have stagnated, become inward-looking and preoccupied with such trivia as hierarchial status and over-formalised manners. It is effeteness of this sort that I fear might result from the exhaustion of discovery. However, that situation will differ from any past static society in a number of ways. The past society that lapsed into stagnation was likely to be over-run or eliminated by more vigorous cultures, thereby continually selecting dynamic cultures by a process of natural selection; but in the state I envisage there will be no "barbarians left beyond the walls".

"The egg timer is pinging. The toaster is popping. The coffeepot is perking. Is this it, Alice? Is this the great American dream?"

(Drawing by H. Martin; © 1973 The New Yorker Magazine, Inc.)

On the other hand it is possible that with "total knowledge" we will know how to avert stagnation. After all, it may not be impossible to keep a person in wonder on his journey from the cradle to the grave, even if we do know all the answers. Perhaps the chief problem—as with so many of our current ills—is the rate at which it will come upon us.

As with oil and mineral resources, so we are going through our resources of discovery at a prodigal rate. Instead of innovating at a rate that taxes our civilisation to—perhaps beyond—its limits a much slower rate would suffice to keep our culture living (an interesting argument for cutting back on science expenditure). But we should like to know the consequences of "total knowledge" and to what extent cultures require the stimulus of discovery. This leads to a paradox: we should decrease our research effort to conserve our reserves of discovery; but to be able to anticipate the consequence of exhaustion of discovery we need more understanding of ourselves and our cultures.

34
WHAT WE MUST DO

JOHN PLATT

There is only one crisis in the world. It is the crisis of transformation. The trouble is that it is now coming upon us as a storm of crisis problems from every direction. But if we look quantitatively at the course of our changes in this century, we can see immediately why the problems are building up so rapidly at this time, and we will see that it has now become urgent for us to mobilize all our intelligence to solve these problems if we are to keep from killing ourselves in the next few years.

The essence of the matter is that the human race is on a steeply rising "S-curve" of change. We are undergoing a great historical transition to new levels of technological power all over the world. We all know about these changes, but we do not often stop to realize how large they are in orders of magnitude, or how rapid and enormous compared to all previous changes in history. In the last century, we have increased our speeds of communication by a factor of 10^7,* our speeds of travel by 10^2; our speeds of data handling by 10^6; our energy resources by 10^3; our power of weapons by 10^6; our ability to control diseases by something like 10^2; and our rate of population growth to 10^3 times what it was a few thousand years ago.

Could anyone suppose that human relations around the world would not be affected to their very roots by such changes? Within the last 25 years, the Western world has moved into an age of jet planes, missiles and satellites, nuclear power and nuclear terror. We have acquired computers and automation, a service and leisure economy, superhighways, superagriculture, supermedicine, mass higher education, universal TV, oral contraceptives, environmental pollution, and urban crises. The rest of the world is also moving rapidly and may catch up with all these powers and problems within a very short time. It is hardly surprising that young people under 30, who have grown up

From *Science 166*: 1115, 1969

*10^7 = ten million; 10^6 = one million; 10^3 = one thousand; 10^2 = one hundred (Ed.)

familiar with these things from childhood, have developed very different expectations and concerns from the older generation that grew up in another world.

What many people do not realize is that many of these technological changes are now approaching certain natural limits. The "S-curve" is beginning to level off. We may never have faster communications or more TV or larger weapons or a higher level of danger than we have now. This means that if we could learn how to manage these new powers and problems in the next few years without killing ourselves by our obsolete structures and behavior, we might be able to create new and more effective social structures that would last for many generations. We might be able to move into that new world of abundance and diversity and well-being for all mankind which technology has now made possible.

The trouble is that we may not survive these next few years. The human race today is like a rocket on a launching pad. We have been building up to this moment of takeoff for a long time, and if we can get safely through the takeoff period, we may fly on a new and exciting course for a long time to come. But at this moment, as the powerful new engines are fired, their thrust and roar shakes and stresses every part of the ship and may cause the whole thing to blow up before we can steer it on its way. Our problem today is to harness and direct these tremendous new forces through this dangerous transition period to the new world instead of to destruction. But unless we can do this, the rapidly increasing strains and crises of the next decade may kill us all. They will make the last 20 years look like a peaceful interlude.

THE NEXT 10 YEARS

Several types of crisis may reach the point of explosion in the next 10 years: nuclear escalation, famine, participatory crises, racial crises, and what have been called the crises of administrative legitimacy. It is worth singling out two or three of these to see how imminent and dangerous they are, so that we can fully realize how very little time we have for preventing or controlling them.

Take the problem of nuclear war, for example. A few years ago, Leo Szilard estimated the "half-life' of the human race* with respect to nuclear escalation as being between 10 and 20 years. His reasoning then is still valid now. As long as we continue to have no adequate stabilizing peace-keeping structures for the world, we continue to live under the daily threat not only of local wars but of nuclear escalation with overkill and megatonnage enough to destroy all life on earth. Every year or two there is a confrontation between nuclear powers—Korea, Laos, Berlin, Suez, Quemoy, Cuba, Vietnam, and the rest. MacArthur wanted to use nuclear weapons in Korea; and in the Cuban missile crisis, John Kennedy is said to have estimated the probability of a nuclear exchange as about 25 percent.

The danger is not so much that of the unexpected, such as a radar error or even a new nuclear dictator, as it is that our present systems will work exactly as planned!—from border testing, strategic gambles, threat and counterthreat, all the way up to that "second-strike capability" that is already aimed, armed, and triggered to wipe out hundreds of millions of people in a 3-hour duel!

What is the probability of this in the average incident? 10 percent? 5 percent? There is no average incident. But it is easy to see that five or ten more such confrontations in this game of "nuclear roulette" might indeed give us only a 50-50 chance of living until 1980 or 1990. This is a shorter life expectancy than people have ever had in the world before. All our medical increases in length of life are meaningless, as long as our nuclear lifetime is so short.

Many agricultural experts also think that within the next decade the great famines will begin, with deaths that may reach 100 million people in densely populated countries like India and China. Some contradict this, claiming that the remarkable new grains and new agricultural methods introduced . . . in Southeast Asia may now be able to keep the food supply ahead of population growth. But others think that the reeducation of farmers and consumers to use the new grains cannot proceed fast enough to make a difference.

But if famine does come, it is clear that it will be catastrophic. Besides the direct

*The length of time during which the probability of human existence would decrease to 50 percent of what it was before (Ed.)

human suffering, it will further increase our international instabilities, with food riots, troops called out, governments falling, and international interventions that will change the whole political map of the world. It could make Vietnam look like a popgun.

In addition, the next decade is likely to see continued crises of legitimacy of all our overloaded administrations, from universities and unions to cities and national governments. Everywhere there is protest and refusal to accept the solutions handed down by some central elite. The student revolutions circle the globe. Suburbs protest as well as ghettoes, Right as well as Left. There are many new sources of collision and protest, but it is clear that the general problem is in large part structural rather than political. Our traditional methods of election and management no longer give administrations the skill and capacity they need to handle their complex new burdens and decisions. They become swollen, unresponsive—and repudiated. Every day now some distinguished administrator is pressured out of office by protesting constituents.*

In spite of the violence of some of these confrontations, this may seem like a trivial problem compared to war or famine—until we realize the dangerous effects of these instabilities on the stability of the whole system. In a nuclear crisis or in any of our other crises today, administrators or negotiators may often work out some basis of agreement between conflicting groups or nations, only to find themselves rejected by their people on one or both sides, who are then left with no mechanism except to escalate their battles further.

THE CRISIS OF CRISES

What finally makes all of our crises still more dangerous is that they are now coming on top of each other. Most administrations are able to endure or even enjoy an occasional crisis, with everyone working late together and getting a new sense of importance and unity. What they are not prepared to deal with are multiple crises, a crisis of crises all at one time. This is what happened in New York City in 1968 when the Ocean Hill-Brownsville teacher and race strike was combined with a police strike, on top of a garbage strike, on top of a longshoremen's strike, all within a few days of each other.

When something like this happens, the staffs get jumpy with smoke and coffee and alcohol, the mediators become exhausted, and the administrators find themselves running two crises behind. Every problem may escalate because those involved no longer have time to think straight. What would have happened in the Cuban missile crisis if the East Coast power blackout had occurred by accident that same day? Or if the "hot line" between Washington and Moscow had gone dead? There might have been hours of misinterpretation, and some fatally different decisions.

I think this multiplication of domestic and international crises today will shorten that short half-life. In the continued absence of better ways of heading off these multiple crises, our half-life may no longer be 10 or 20 years, but more like 5 to 10 years, or less. We may have even less than a 50–50 chance of living until 1980.

This statement may seem uncertain and excessively dramatic. But is there any scientist who would make a much more optimistic estimate after considering all the different sources of danger and how they are increasing? The shortness of the time is due to the exponential and multiplying character of our problems and not to what particular numbers or guesses we put in. Anyone who feels more hopeful about getting past the nightmares of the 1970's has only to look beyond them to the monsters of pollution and population rising up in the 1980's and 1990's. Whether we have 10 years or more like 20 or 30, unless we systematically find new large-scale solutions, we are in the gravest danger of destroying our society, our world, and ourselves in any of a number of different ways well before the end of this century. Many futurologists who have predicted what the world will be like in the year 2000 have neglected to tell us that.

Nevertheless the real reason for trying to make rational estimates of these deadlines is not because of their shock value but because they give us at least a rough idea of how much time we may have for finding and mounting some large-scale solutions. The time is short but, as we shall see, it is not too short to give us a chance that something can be done, if we begin immediately.

*This was written some five years before the demise of the Nixon administration. (Ed.)

From this point, there is no place to go but up. Human predictions are always conditional. The future always depends on what we do and can be made worse or better by stupid or intelligent action. To change our earlier analogy, today we are like men coming out of a coal mine who suddenly begin to hear the rock rumbling, but who have also begun to see a little square of light at the end of the tunnel. Against this background, I am an optimist—in that I want to insist that there is a square of light and that it is worth trying to get to. I think what we must do is to start running as fast as possible toward that light, working to increase the probability of our survival through the next decade by some measurable amount.

For the light at the end of the tunnel is very bright indeed. If we can only devise new mechanisms to help us survive this round of terrible crises, we have a chance of moving into a new world of incredible potentialities for all mankind. But if we cannot get through this next decade, we may never reach it.

TASK FORCES FOR SOCIAL RESEARCH AND DEVELOPMENT

What can we do? I think that nothing less than the application of the full intelligence of our society is likely to be adequate. These problems will require the humane and constructive efforts of everyone involved. But I think they will also require something very similar to the mobilization of scientists for solving crisis problems in wartime. I believe we are going to need large numbers of scientists forming something like research teams or task forces for social research and development. We need full-time interdisciplinary teams combining men of different specialties, natural scientists, social scientists, doctors, engineers, teachers, lawyers, and many other trained and inventive minds, who can put together our stores of knowledge and powerful new ideas into improved technical methods, organizational designs, or "social inventions" that have a chance of being adopted soon enough and widely enough to be effective. Even a great mobilization of scientists may not be enough. There is no guarantee that these problems can be solved, or solved in time, no matter what we do. But for problems of this scale and urgency, this kind of focusing of our brains and knowledge may be the only chance we have.

Scientists, of course, are not the only ones who can make contributions. Millions of citizens, business and labor leaders, city and government officials, and workers in existing agencies, are already doing all they can to solve these problems. No scientific innovation will be effective without extensive advice and help from all these groups.

But it is the new science and technology that have made our problems so immense and intractable. Technology did not create human conflicts and inequities, but it has made them unendurable. And where science and technology have expanded the problems in this way, it may be only more scientific understanding and better technology that can carry us past them. The cure for the pollution of the rivers by detergents is the use of nonpolluting detergents. The cure for bad management designs is better management designs.

Also, in many of these areas, there are few people outside the research community who have the basic knowledge necessary for radically new solutions. In our great biological problems, it is the new ideas from cell biology and ecology that may be crucial. In our social-organizational problems, it may be the new theories of organization and management and behavior theory and game theory that offer the only hope. Scientific research and development groups of some kind may be the only effective mechanism by which many of these new ideas can be converted into practical invention and action.

The time scale on which such task forces would have to operate is very different from what is usual in science. In the past, most scientists have tended to work on something like a 30-year time scale, hoping that their careful studies would fit into some great intellectual synthesis that might be years away. Of course when they become politically concerned, they begin to work on something more like a 3-month time scale, collecting signatures or trying to persuade the government to start or stop some program.

But 30 years is too long, and 3 months is too short, to cope with the major crises that might destroy us in the next 10 years. Our urgent problems now are more like wartime problems, where we need to work as rapidly as is consistent with large-scale effectiveness. We need to think rather in terms of a 3-year time scale—or more broadly, a 1- to 5-year time scale. In World War II, the ten thousand scientists who were mobilized for war research knew they did not have 30 years, or even 10 years, to come up with answers. But they did have time for the new research, design, and construction that brought sonar and radar and atomic energy to operational effectiveness within 1 to 4 years. Today we need the same large-scale mobilization for innovation and action and the same sense of constructive urgency.

PRIORITIES: A CRISIS INTENSITY CHART

In any such enterprise, it is most important to be clear about which problems are the real priority problems. To get this straight, it is valuable to try to separate the different problem areas according to some measures of their magnitude and urgency. A possible classification of this kind is shown in Tables 1 and 2. In these tables I have tried to rank a number of present or potential problems or crises, vertically, according to an estimate of their order of intensity or "seriousness," and horizontally, by a rough estimate of their time to reach climactic importance. Table 1 is such a classification for the United States for the next 1 to 5 years, the next 5 to 20 years, and the next 20 to 50 years. Table 2 is a similar classification for world problems and crises.

The successive rows indicate something like order-of-magnitude differences in the intensity of the crises, as estimated by a rough product of the size of population that might be hurt or affected, multiplied by some estimated average effect in the disruption of their lives. Thus the first row corresponds to total or near-total annihilation; the second row, to great destruction or change affecting everybody; the third row, to a lower tension affecting a smaller part of the population or a smaller part of everyone's life, and so on.

Informed men might easily disagree about one row up or down in intensity, or one column left or right in the time scales, but these order-of-magnitude differences are already so great that it would be surprising to find much larger disagreements. Clearly, an important initial step in any serious problem study would be to refine such estimates.

In both tables, the one crisis that must be ranked at the top in total danger and imminence is, of course, the danger of large-scale or total annihilation by nuclear escalation or by radiological-chemical-biological-warfare (RCBW). This kind of crisis will continue through both the 1- to 5-year time period and the 5- to 20-year period as Crisis Number 1, unless and until we get a safer peace-keeping arrangement. But in the 20- to 50-year column, following the reasoning already given, I think we must simply put a big "✠" at this level, on the grounds that the peace-keeping stabilization problem will either be solved by that time or we will probably be dead.

At the second level, the 1- to 5-year period may not be a period of great destruction (except nuclear) in either the United States or the world. But the problems at this level are building up, and within the 5- to 20-year period, many scientists fear the destruction of our whole biological and ecological balance in the United States by mismanagement or pollution. Others fear political catastrophe within this period, as a result of participatory confrontations or backlash or even dictatorship, if our divisive social and structural problems are not solved before that time.

On a world scale in this period, famine and ecological catastrophe head the list of destructive problems. We will come back later to the items in the 20- to 50-year column.

The third level of crisis problems in the United States includes those that are already upon us: administrative management of communities and cities, slums, participatory democracy, and racial conflict. In the 5- to 20-year period, the problems of pollution and poverty or major failures of law and justice could escalate to this level of tension if they are not solved. The last column is left blank because secondary events and second-order effects will interfere seriously with any attempt to make longer-range predictions at these lower levels.

The items in the lower part of the tables are not intended to be exhaustive. Some are common headline problems which are included simply to show how they might rank quantitatively in this kind of comparison. Anyone concerned with any of them will find it a useful exercise to estimate for himself their order of seriousness, in terms of the number of people they actually affect and

TABLE 1. CLASSIFICATION OF PROBLEMS AND CRISES BY ESTIMATED TIME AND INTENSITY (UNITED STATES).

Grade	Estimated crisis intensity (number affected × degree of effect)	Crisis or problem	Estimated time to crisis*		
			1 to 5 Years	5 to 20 Years	20 to 50 Years
1.		Total annihilation	Nuclear or RCBW escalation	Nuclear or RCBW escalation	✢ (Solved or dead)
2.	10^8	Great destruction or change (physical, biological, or political)	(Too soon)	Participatory democracy Ecological balance ↑	Political theory and economic structure Population planning Patterns of living Education Communications Integrative philosophy ↑
3.	10^7	Widespread almost unbearable tension	Administrative management Slums Participatory democracy Racial conflict	Pollution Poverty Law and justice	↑ ?
4.	10^6	Large-scale distress	Transportation Neighborhood ugliness Crime	Communications gap	?
5.	10^5	Tension producing responsive change	Cancer and heart Smoking and drugs Artificial organs Accidents Sonic boom Water supply Marine resources Privacy on computers	Educational inadequacy	?
6.		Other problems—important, but adequately researched	Military R & D New educational methods Mental illness Fusion power	Military R & D	
7.		Exaggerated dangers and hopes	Mind control Heart transplants Definition of death	Sperm banks Freezing bodies Unemployment from automation	Eugenics
8.		Noncrisis problems being "overstudied"	Man in space Most basic science		

*If no major effort is made at anticipatory solution.

TABLE 2. CLASSIFICATION OF PROBLEMS AND CRISES BY ESTIMATED TIME AND INTENSITY (WORLD).

Grade	Estimated crisis intensity (number affected × degree of effect)	Crisis or problem	Estimated time to crisis*		
			1 to 5 years	*5 to 20 years*	*20 to 50 years*
1.	10^{10}	Total annihilation	Nuclear or RBCW escalation	Nuclear or RCBW escalation	✢ (Solved or dead)
2.	10^{9}	Great destruction or change (physical, biological, or political)	(Too soon)	Famines Ecological balance Development failures Local wars Rich-poor gap	Economic structure and political theory Population and ecological balance Patterns of living Universal education Communications-integration Management of world Integrative philosophy
3.	10^{8}	Widespread almost unbearable tension	Administrative management Need for participation Group and racial conflict Poverty-rising expectations Environmental degradation	Poverty Pollution Racial wars Political rigidity Strong dictatorships	?
4.	10^{7}	Large-scale distress	Transportation Diseases Loss of old cultures	Housing Education Independence of big powers Communications gap	?
5.	10^{6}	Tension producing responsive change	Regional organization Water supplies	?	?
6.		Other problems—important, but adequately researched	Technical development design Intelligent monetary design		
7.		Exaggerated dangers and hopes			Eugenics Melting of ice caps
8.		Noncrisis problems being "overstudied"	Man in space Most basic science		

*If no major effort is made at anticipatory solution.

the average distress they cause. Transportation problems and neighborhood ugliness, for example, are listed as grade 4 problems in the United States because they depress the lives of tens of millions for 1 or 2 hours every day. Violent crime may affect a corresponding number every year or two. These evils are not negligible, and they are worth the efforts of enormous numbers of people to cure them and to keep them cured—but on the other hand, they will not destroy our society.

The grade 5 crises are those where the hue and cry has been raised and where responsive changes of some kind are already under way. Cancer goes here, along with problems like auto safety and an adequate water supply. This is not to say that we have solved the problem of cancer, but rather that good people are working on it and are making as much progress as we could expect from anyone. (At this level of social intensity, it should be kept in mind that there are also positive opportunities for research, such as the automation of clinical biochemistry or the invention of new channels of personal communication, which might affect the 20-year future as greatly as the new drugs and solid state devices of 20 years ago have begun to affect the present.)

WHERE THE SCIENTISTS ARE

Below grade 5, three less quantitative categories are listed, where the scientists begin to outnumber the problems. Grade 6 consists of problems that many people believe to be important but that are adequately researched at the present time. Military R & D belongs in this category. Our huge military establishment creates many social problems, both of national priority and international stability, but even in its own terms, war research, which engrosses hundreds of thousands of scientists and engineers, is being taken care of generously. Likewise, fusion power is being studied at the $100-million level, though even if we had it tomorrow, it would scarcely change our rates of application of nuclear energy in generating more electric power for the world.

Grade 7 contains the exaggerated problems which are being talked about or worked on out of all proportion to their true importance, such as heart transplants, which can never affect more than a few thousands of people out of the billions in the world. It is sad to note that the symposia on "social implications of science" at many national scientific meetings are often on the problems of grade 7.

In the last category, grade 8, are two subjects which I am sorry to say I must call "over-studied," at least with respect to the real crisis problems today. The Man in Space flights to the moon and back are the most beautiful technical achievements of man, but they are not urgent except for national display, and they absorb tens of thousands of our most ingenious technical brains.

And in the "overstudied" list I have begun to think we must now put most of our basic* science. This is a hard conclusion, because all of science is so important in the long run and because it is still so small compared, say, to advertising or the tobacco industry. But basic scientific thinking is a scarce resource. In a national emergency, we would suddenly find that a host of our scientific problems could be postponed for several years in favor of more urgent research. Should not our total human emergency make the same claims? Long-range science is useless unless we survive to use it. Tens of thousands of our best trained minds may now be needed for something more important than "science as usual."

The arrows at level 2 in the tables are intended to indicate that problems may escalate to a higher level of crisis in the next time period if they are not solved. The arrows toward level 2 in the last columns of both tables show the escalation of all our problems upward to some general reconstruction in the 20- to 50-year time period, if we survive. Probably no human institution will continue unchanged for another 50 years, because they will all be changed by the crises if they are not changed in advance to prevent them. There will surely be widespread rearrangements in all our ways of life everywhere, from our patterns of society to our whole philosophy of man. Will they be more humane, or less? Will the world come to resemble a diverse and open humanist democracy? Or Orwell's *1984?* Or a postnuclear desert with its scientists hanged? It is our acts of commitment and leadership in

*As opposed to applied (Ed.)

the next few months and years that will decide.

MOBILIZING SCIENTISTS

It is a unique experience for us to have peacetime problems, or technical problems which are not industrial problems, on such a scale. We do not know quite where to start, and there is no mechanism yet for generating ideas systematically or paying teams to turn them into successful solutions.

But the comparison with wartime research and development may not be inappropriate. Perhaps the antisubmarine warfare work or the atomic energy project of the 1940's provides the closest parallels to what we must do in terms of the novelty, scale, and urgency of the problems, the initiative needed, and the kind of large success that has to be achieved. In the antisubmarine campaign, Blackett assembled a few scientists and other ingenious minds in his "back room," and within a few months they had worked out the "operations analysis" that made an order-of-magnitude difference in the success of the campaign. In the atomic energy work, scientists started off with extracurricular research, formed a central committee to channel their secret communications, and then studied the possible solutions for some time before they went to the government for large-scale support for the great development laboratories and production plants.

Fortunately, work on our crisis problems today would not require secrecy. Our great problems today are all beginning to be world problems, and scientists from many countries would have important insights to contribute.

Probably the first step in crisis studies now should be the organization of intense technical discussion and education groups in every laboratory. Promising lines of interest could then lead to the setting up of part-time or full-time studies and teams and coordinating committees. Administrators and boards of directors might find active crisis research important to their own organizations in many cases. Several foundations and federal agencies already have inhouse research and make outside grants in many of these crisis areas, and they would be important initial sources of support.

But the step that will probably be required in a short time is the creation of whole new centers, perhaps comparable to Los Alamos or the RAND Corporation, where interdisciplinary groups can be assembled to work full-time on solutions to these crisis problems. Many different kinds of centers will eventually be necessary, including research centers, development centers, training centers, and even production centers for new sociotechnical inventions. The problems of our time—the $100-billion food problem or the $100-billion arms control problem—are no smaller than World War II in scale and importance, and it would be absurd to think that a few academic research teams or a few agency laboratories could do the job.

SOCIAL INVENTIONS

The thing that discourages many scientists—even social scientists—from thinking in these research-and-development terms is their failure to realize that there are such things as social inventions and that they can have large-scale effects in a surprisingly short time. A recent study with Karl Deutsch has examined some 40 of the great achievements in social science in this century, to see where they were made and by whom and how long they took to become effective. They include developments such as the following:

- Keynesian economics
- Opinion polls and statistical sampling
- Input-output economics
- Operations analysis
- Information theory and feedback theory
- Theory of games and economic behavior
- Operant conditioning and programmed learning
- Planned programming and budgeting (PPB)
- Non–zero-sum game theory

Many of these have made remarkable differences within just a few years in our ability to handle social problems or management problems. The opinion poll became a national necessity within a single election period. The theory of games, published in 1946, had become an important component of American strategic thinking by RAND and the Defense Department by 1953, in spite of the limitation of the theory at that time to zero-sum games, with their dangerous bluffing and "brinksmanship." Today, within less than a decade, the PPB management technique is sweeping through every large organization.

This list is particularly interesting because it shows how much can be done outside official government agencies when inventive men put their brains together. Most of the achievements were the work of teams of two or more men, almost all of them located in intellectual centers such as Princeton or the two Cambridges.

The list might be extended by adding commercial social inventions with rapid and widespread effects, like credit cards. And sociotechnical inventions, like computers and automation or like oral contraceptives, which were in widespread use within 10 years after they were developed. In addition, there are political innovations like the New Deal, which made great changes in our economic life within 4 years, and the pay-as-you-go income tax, which transformed federal taxing power within 2 years.

On the international scene, the Peace Corps, the "hot line," the Test-Ban Treaty, the Antarctic Treaty, and the Nonproliferation Treaty were all implemented within 2 to 10 years after their initial proposal. These are only small contributions, a tiny patchwork part of the basic international stabilization system that is needed, but they show that the time to adopt new structural designs may be surprisingly short. Our cliches about "social lag" are very misleading. Over half of the major social innovations since 1940 were adopted or had widespread social effects within less than 12 years—a time as short as, or shorter than, the average time for adoption of technological innovations.

AREAS FOR TASK FORCES

Is it possible to create more of these social inventions systematically to deal with our present crisis problems? I think it is. It may be worth listing a few specific areas where new task forces might start.

1) *Peace-keeping mechanisms and feedback stabilization.* Our various nuclear treaties are a beginning. But how about a technical group that sits down and thinks about the whole range of possible and impossible stabilization and peace-keeping mechanisms? Stabilization feedback-design might be a complex modern counterpart of the "checks and balances" used in designing the constitutional structure of the United States 200 years ago. With our new knowledge today about feedbacks, group behavior, and game theory, it ought to be possible to design more complex and even more successful structures.

Some peace-keeping mechanisms that might be hard to adopt today could still be worked out and tested and publicized, awaiting a more favorable moment. Sometimes the very existence of new possibilities can change the atmosphere. Sometimes, in a crisis, men may finally be willing to try out new ways and may find some previously prepared plan of enormous help.

2) *Biotechnology.* Humanity must feed and care for the children who are already in the world, even while we try to level off the further population explosion that makes this so difficult. Some novel proposals, such as food from coal, or genetic copying of champion animals, or still simpler contraceptive methods, could possibly have large-scale effects on human welfare within 10 to 15 years. New chemical, statistical, and management methods for measuring and maintaining the ecological balance could be of very great importance.

3) *Game theory.* As we have seen, zero-sum game theory has not been too academic to be used for national strategy and policy analysis. Unfortunately, in zero-sum games, what I win, you lose, and what you win, I lose. This may be the way poker works, but it is not the way the world works. We are collectively in a non-zero-sum game in which we will all lose together in nuclear holocaust or race conflict or economic nationalism, or all win together in survival and prosperity. Some of the many variations of non–zero-sum game theory, applied to group conflict and cooperation, might show us profitable new approaches to replace our sterile and dangerous confrontation strategies.

4) *Psychological* and *social theories.* Many teams are needed to explore in detail and in practice how the powerful new ideas of behavior theory and the new ideas of responsive living might be used to improve family life or community and management structures. New ideas of information handling and management theory need to be turned into practical recipes for reducing the daily frustrations of small businesses,

schools, hospitals, churches, and town meetings. New economic inventions are needed, such as urban development corporations. A deeper systems analysis is urgently needed to see if there is not some practical way to separate full employment from inflation. Inflation pinches the poor, increases labor-management disputes, and multiplies all our domestic conflicts and our sense of despair.

5) *Social indicators.* We need new social indicators, like the cost-of-living index, for measuring a thousand social goods and evils. Good indicators can have great "multiplier effects" in helping to maximize our welfare and minimize our ills. Engineers and physical scientists working with social scientists might come up with ingenious new methods of measuring many of these important but elusive parameters.

6) *Channels of effectiveness.* Detailed case studies of the reasons for success or failure of various social inventions could also have a large multiplier effect. Handbooks showing what channels or methods are now most effective for different small-scale and large-scale social problems would be of immense value.

The list could go on and on. In fact, each study group will have its own pet projects. Why not? Society is at least as complex as, say, an automobile with its several thousand parts. It will probably require as many research-and-development teams as the auto industry [has] in order to explore all the inventions it needs to solve its problems. But it is clear that there are many areas of great potential crying out for brilliant minds and brilliant teams to get to work on them.

FUTURE SATISFACTIONS AND PRESENT SOLUTIONS

This is an enormous program. But there is nothing impossible about mounting and financing it, if we, as concerned men, go into it with commitment and leadership. Yes, here will be a need for money and power to overcome organizational difficulties and vested interests. But it is worth remembering that the only real source of power in the world is the gap between what is and what might be. Why else do men work and save and plan? If there is some future increase in human satisfaction that we can point to and realistically anticipate, men will be willing to pay something for it and invest in it in the hope of that return. In economics, they pay with money; in politics, with their votes and time and sometimes with their jail sentences and their lives.

Social change, peaceful or turbulent, is powered by "what might be." This means that for peaceful change, to get over some impossible barrier of unresponsiveness or complexity or group conflict, what is needed is an inventive man or group—a "social entrepreneur"—who can connect the pieces and show how to turn the advantage of "what might be" into some present advantage for every participating party. To get toll roads, when highways were hopeless, a legislative-corporation mechanism was invented that turned the future need into present profits for construction workers and bondholders and continuing profitability for the state and all the drivers.

This principle of broad-payoff anticipatory design has guided many successful social plans. Regular task forces using systems analysis to find payoffs over the barriers might give us such successful solutions much more often. The new world that could lie ahead, with its blocks and malfunctions removed, would be fantastically wealthy. It seems almost certain that there must be many systematic ways for intelligence to convert that large payoff into the profitable solution of our present problems.

The only possible conclusion is a call to action. Who will commit himself to this kind of search for more ingenious and fundamental solutions? Who will begin to assemble the research teams and the funds? Who will begin to create those full-time interdisciplinary centers that will be necessary for testing detailed designs and turning them into effective applications?

The task is clear. The task is huge. The time is horribly short. In the past, we have had science for intellectual pleasure, and science for the control of nature. We have had science for war. But today, the whole human experiment may hang on the question of how fast we now press the development of science for survival.

35

SCIENCE SERVES SOCIETY*

LEE A. DuBRIDGE

No matter how much a university professor or president has dealt with the government from the outside, he begins to really learn about the problems of science and government only when he is on the inside. And no matter how much he may learn during his first 3 months in office, the main thing he realizes is that he really knows very little. There are, I think, three simple reasons for this: (i) Science is very complicated; (ii) government is very complicated; and (iii) when you multiply the complications of science by those of government, you get a very large number of complications indeed.

We are all familiar with the complications of science and of scientists. We know the enormous spectrum of activities covered by the phrase "science and technology." We know that there are several hundred thousand people in the United States working on scientific and engineering projects, and each one, I am sure, thinks that his field of interest is just about the most important one there is.

This is one reason why scientists and engineers are complicated. We are all devoted to our own pursuits and are unanimous in our belief that science and engineering are important to the country and to the world. However, that is about the only thing we are unanimous about. We are not unanimous about the relative importance of various fields of basic science; we are not unanimous about the relative merits of basic and applied science; and we are surely not unanimous on priorities in the field of applied science—about the uses to which scientific knowledge should be put. Thus, when anyone seeks to find the opinion of the scientific world on a particular public issue—whether it be the ABM, the SST, the space program, government support of research, or many others—the layman is understandably astounded at the wide variety of very strong

*This article is adapted from an address given to the members of the National Academy of Sciences on 29 April 1969 in Washington, D.C.

From *Science 164*: 1137, 1969.

opinions that may be offered. "Can't the scientists make up their minds?" we hear it asked.

Now the spectrum of opinion among scientists is quite understandable. Public issues of the sort which I have mentioned are not purely scientific issues. They are issues in which science and technology constitute only one component. Other components may involve fiscal affairs, political matters, social conditions, international relations, inflation, taxes, and even moral judgments. In these areas we are all laymen and in these areas we find ourselves in the strange situation (strange to us as scientists) of being asked to render judgments on the basis of data which are inexact, incomplete, often conflicting, and surely not obtained under the rigid conditions of the controlled experiment which we take for granted in our scientific laboratories.

Thus while scientists may well come to substantial agreement on such questions as the validity of the special theory of relativity, or of quantum mechanics, or the structure of protein or DNA molecules, we find ourselves in wide disagreement on whether the government should invest funds in enterprises which involve both technical and nontechnical factors. We also disagree on the priorities which the nation should assign in funding various fields of applied or pure science; for example, space, oceanography, astronomy, high energy physics, microbiology, urban development, and others.

This leads me to my second point, namely, the government is complicated. I have personally known for a long time that the U.S. government is a complex enterprise. The last 3 months* have revealed complexities of which I was previously unaware—or only dimly aware. I shall discuss only one of the government's many complexities: its involvement in pure and applied science. It was once true—say 40 years ago—that the government's involvement in science was quite simple: it didn't exist. Actually, that is not quite true. We did have the National Bureau of Standards, the Smithsonian Institution, the Geological Survey, the Naval Research Laboratory, and a few other scientific establishments—often excellent and important in their own fields but small enough *in toto* to attract very little attention from the Congress or the public. There were not too many scientists and they did not spend much federal money. So no one had to pay attention to them.

We are all aware of how that situation has changed. Now there are many scientists and they spend quite a lot of money. Every old-line department of government, plus a great array of new departments and agencies, are now heavily involved in science and technology. Hundreds of thousands of scientists and engineers now work for the government, directly or through its contractors. The government annually spends some $2 billion for research and $15 billion for development. And no one, in or out of government, would assert that there are not some difficult problems and complexities involved in this enterprise.

The government does not spend billions of dollars a year on science and technology without getting involved with those complex people called scientists and engineers. And when the complex array of people and agencies in government get mixed up with the complex scientific community, the result is almost bound to be utter confusion. And, indeed, confusion is what we find.

And yet, if one looks a little deeper one can find that within the dense cloud of confusion there are areas of brightness, areas of directed and purposeful motion, areas where chaos has given way—or is giving way—to order.

SCIENTIFIC ACHIEVEMENTS

Let us then look at a few things on which we can all agree—things on which we as scientists agree and things on which most educated laymen will also agree. The first is one we often forget or at least neglect to emphasize: namely, that during the past 300 years the worldwide community of scientists has built up an astounding, indeed a miraculous, structure of accurate and verifiable facts and principles about the physical universe including the universe of living things. I never can resist a sense of awe and wonder as I reflect on the things we all know today which no one knew 100 years ago, or 50 years ago, or even 5 years ago. In fact, somewhere, someone has very probably learned today something that no one knew yesterday. Surely the scientific enterprise, judged in

*as Director of the Office of Science and Technology [Ed.]

terms of attaining its primary objective—the accumulation of knowledge and understanding—has been the most brilliantly successful enterprise in human history. We are proud to be scientists because we are proud of the opportunity to be able to contribute, if only a little bit, to this magnificent and rapidly growing structure of knowledge and understanding.

But at once we face a curious ambiguity in our feelings. While we take pride in what we have learned, we are, at the same time, humble as we face the things we do not know or understand. We are still in the position of Isaac Newton who beheld with awe the seas of ignorance which extended beyond his beachhead of knowledge. How we can, at the same time, be proud of our collective and cumulated knowledge and distressed at our ignorance is something which many laymen find hard to understand. But there it is, and it is a basic fact in our lives. Our urge to push ahead on the frontiers of ignorance quickly overcomes any temptation we might have to sit back and compliment ourselves on what we have learned. Yet, I suggest that now and then we should do that too.

There is another fact on which we can all agree: that, in spite of some doubts and reservations here and there, we are certain that this great body of scientific knowledge has been of enormous benefit to the human race. Since some people are questioning this today, we should take another hard look at the balance sheet. Even though we can spot some red figures here and there, we must conclude that the assets far exceed the liabilities. To prove this we need only look back 100 years and ask how men lived and thought and worked then as compared to today. The material changes are obvious. In those areas of the world where scientific knowledge has really been put to use, man's health, comfort, wealth, and welfare have been enormously enhanced. A substantial segment of the human race—even if not all of it—has lifted itself to heights of comfort, leisure, and affluence undreamed of, and yet desperately desired, a century ago. Indeed, as we contemplate the contrasts in living standards between various parts of the world today, we speak immediately of the technological gap. And that is just what it is. In some parts of the world, scientific knowledge, through technology, has been brought to the benefit of the people; in other parts, it has not. Whether we look around the world or look backward in time, the vast benefits of scientific knowledge are clearly evident.

SOCIAL PROGRESS AIDED BY SCIENCE

But we need not confine attention to material things. Many thoughtful people have noted that it is surely no accident that the rise of science and technology has been paralleled in time and in place with the rise of democratic governments—governments based on a recognition of the dignity and worth of every human being. Men who can understand the universe and life cannot easily tolerate human misery and injustice. When knowledge replaces superstition, men think and act differently. Once, man in his ignorance looked upon poverty and disease as inevitable—and possibly even ordained by the gods. Today he knows these things can be conquered and he is impatient to get on with the job.

In fact, I firmly believe that the ideals of men for the betterment of the lot of human beings are higher today than ever before in history—and that the advance of science and technology, which has helped us to achieve many ideals, has also enhanced our determination to move on to the attainment of even higher ideals. As we eliminate one cause of human suffering, we yearn to eliminate them all, and we become ever more impatient with our slow progress—even though the rate of progress is in fact accelerating. For example, air pollution in many of our cities is far less today than it was 30 years ago when the pall of soft-coal smoke used to choke us and turn day into night. But as technology, based on scientific knowledge, abated that source of pollution, we lifted our sights and are now determined to eliminate all sources—both old and new. Our ideals have advanced faster than our ability to keep up with them.

This indeed is not an uncommon situation in the world today. The success of our technology in solving some problems has elevated our determination to solve many more human problems more rapidly. There are those who say that science and technology have moved ahead more rapidly than our moral and social standards. There is a case to be made for the reverse: that our humane goals have advanced faster than the ability of our science and technology—plus the ability of our economic, social and political

"But we just don't have the technology to carry it out."

From Sidney Harris, American Scientist.

institutions and skills—to keep up. Every success of technology only seems to widen the gap between what we can do and what we want to do. Thus the pride in our successes is overshadowed by disappointment in our failures to fulfill our rising expectations.

I do not decry this situation; indeed, I applaud it. But I think we should at least recognize it. And we should recognize the corollary: that the scientific community is today a leading influence in advancing our social morality.

To say this is not, of course, to claim that we have been eminently successful. There are many elements in our society who do not share our high ideals for a better world or who do not share our confidence that a better world can be attained. Indeed, we ourselves become often discouraged as we see the gap between the power of our constructive technological skills and the weakness of our economic and political machinery. There are those who will say that the way to close this gap is to weaken our scientific and technological competence. This is pure defeatism. We must, of course, work harder to improve our knowledge and competence in social, economic, and political areas. But as we do so, we will need more than ever a strong science and technology to provide the tools to move ahead toward our goals.

BASIC RESEARCH IS ESSENTIAL

If we agree on this last point then the question is, what do we do about it?

The first thing we must do, I think, is to readdress ourselves to the task of insuring the strength and vitality of basic science, both in this country and throughout the world. Our first responsibility in this direction is obvious: to continue to do good science; to continue to use all the talent and all the resources at our disposal to discover the important secrets of nature which still lie hidden from our view. The American scientific community needs no exhortation from me on this subject. But I would be remiss if I did not reemphasize its basic importance, and the importance of conveying this spirit to all our colleagues and students.

A second task is that we try to speak to the world with one voice on the importance of basic science. Whether we are speaking to our nonscientific colleagues, to administrators in the university or in business and industrial organizations, to representatives of government at any level or to the public at large, we need to speak audibly and forcefully and, if possible, unanimously on this basic point: the discovery of new knowledge is an enterprise of prime importance to the human spirit and to the human condition.

We will not all have the same reasons for expressing this conviction. Some of us will give more emphasis to the cultural values of science; others to the technological values; others to the social or the educational values. All are important; all deserve emphasis. All I suggest is that we do not try to persuade a congressman to support science because it is so much *fun* for the scientists. Fun though it may be to us, we must remember that Congress is not interested in supporting an amusement park for the scientific community. We face a great and difficult task in trying to convince the American people and their elected representatives in Congress that the future of our country, of our people, the future of human beings everywhere, depends in a critical way on the foundation of basic knowledge which we are laying today.

We will not be unanimous about one aspect of this problem, that is, which fields of science are in greatest need of added support. We all know our own field is of great importance; and we all know that our own field is grossly underfunded. Often we may be tempted to argue that certain other fields are overfunded. I hope this temptation can be avoided, at least in our public statements. Our objective should be to increase the total support of basic science. Though we can each present the case for our own field, we should rejoice and not weep if other fields seem, at the moment at least, to be better off.

PRIORITIES IN SCIENCE

This matter of what we call the priorities of various fields of pure or applied science is one of the most difficult and confusing questions which we face. What do we mean by priorities? And even when we decide on priorities, how do we interpret this in terms of private or government effort or budget allocations? To illustrate the difficulty, let me take a concrete case. What do we mean, let us say, when we talk about the relative priority of microbiology compared to high energy physics. (You may choose any two fields you wish.) Do we mean one field has greater importance? If so, importance to whom or to what purpose? Social importance? Importance to human life? To our economy? To the advance of our culture? To the elevation of the human spirit? Or to satisfying the basic urge of human beings to know and to understand? Or to the welfare of scientists? If we confine attention to any one of these goals, we still face a dilemma. Do we mean immediate or long-range importance? Do we mean the specifically foreseeable importance of the results to be attained or to the long-range effects which might be anticipated or imagined? And how does one even foresee or predict the long- or short-range results and applications of basic investigations? We can all think of too many cases of totally unexpected results of research and their wholly unexpected and unforeseeable impact to have any confidence in anyone's prediction that one field of research will surely lead to beneficial results and another one will not.

You can see that the unanswered questions far exceed the number of clear or possible answers—and thus broad and conclusive and universally agreed-upon priority conclusions remain as unreachable as ever.

But suppose, by some magic, we could agree upon a list of priorities among various fields of science, and we could rank them in order: 1, 2, 3, 4, 5 . . . Then what? Shall we assign, say, $100 million to number 1; $90 million to number 2; $80 million to number 3, and so on, until we run out of money and then assign zero to all the rest? Or suppose two fields, say, microbiology and high energy physics, are assigned equal priority. Does that mean they should get equal money? We all know that to do any high energy physics at all we need to have large and expensive accelerators. So even though $100 million might be all that microbiologists could use effectively, that sum will not give us a superhigh energy accelerator. In other words, the funds needed to pursue two areas of science have no necessary relation at all to the priority question.

It is, of course, a little easier to judge relative importance when it comes to applied science for there we are seeking a very specific goal or product whose intrinsic importance we should be able to judge. But do we find the American people or their congressmen unanimous in those judgments? Even if they were, would the answers to the funding questions not still be difficult? Would you put *all* the federal income into priority number one, and nothing into anything else? Or what?

Unanswerable questions! Yet these are the questions with which the executive and legislative branches of our government are struggling every day. The voice of public opinion will be critical in these decisions. Scientists constitute one element of public opinion. Let their voices be heard! Peacefully, I trust! . . .

The members of the Council of the National Academy of Sciences who met with the President on 28 April 1969 heard directly from him his views on this issue. He stated that a nation which devotes exclusive attention to its immediate troubles and problems is bound to decay. Rather, we must look outward and upward in order to lay the foundation for a better future. The advance of scientific knowledge, he said, is an essential enterprise in our society. . . .

FUNDING PROBLEMS

However, to say that the Administration is directing the resources of science, social science, and technology to bear on our social problems is not to say there are not grave difficulties ahead. To put it bluntly, no one knows just how to proceed with this task. We are applying massive funds to temporary palliatives in the form of welfare, relief, alleviation of poverty, food for the hungry, public housing, special educational programs, Medicare, and all the rest. But we do not fully understand the basic problems of our cities or of the poorer rural areas. We do not fully understand how to deal with the problems of improving our society or our environment. As we seek solutions to these problems, we run up against barriers of technology, of economics, of political conflicts, and of inadequate knowledge of what happens to our people and environment as we seek to expand our industrial and agricultural economy, on the one hand, and to make our cities and countryside more liveable, on the other.

While we struggle with immediate and obvious problems requiring large monetary expenditures, we must try at the same time to mount research efforts in which scientists, social and political scientists, and engineers work together to seek basic causes, to develop new technologies, to invent new social and political instrumentalities, to identify and experiment with long-range solutions. Unfortunately, there are not many research centers where such things can be done. There are very few trained people available. The methods and traditions of research which we take for granted in the natural sciences are not so highly developed in these new interdisciplinary areas. Nor is it solely a matter of money. . . . Where are the people? The ideas? The centers of excellence? Even finding a few knowledgeable and devoted people to come into government to staff the R & D operations is proving most difficult. Many people—including myself—believe it is important to have a massive R & D effort in this area, but the fact is that we will have to be content with modest beginnings. If a few more great universities will initiate or accelerate their efforts in research and education in the urban and environmental fields, an enormous contribution would be made.

SCIENCE IS IN POLITICS

I conclude then where I began: the relations between science, technology, government, and the various elements of our society are enormously complex. Science and technology are no longer separable from political and social problems. In these days scientists frequently find themselves engaged in political discussion and activities. When we meet the politicians on their own ground we must not be surprised if they judge us on the basis of our political opinions rather than on the basis of our scientific competence. Whether we like it or not, science is in politics and politics is in science.

Some have said that science is too important to get mixed up in politics. The fact is that today science is too important to stay out of politics. For in our democracy, it is through politics that things get done.

Clearly we all—politicians and scientists—must find ways of adapting ourselves to a new era—an era which began . . . on Hiroshima day in 1945. If we all try, we can accommodate ourselves to the situation.

PROGNOSIS

Laurence E. Strong

As you have seen from the foregoing selections, while scientists may agree that science is an important human activity, they do not necessarily agree on how a scientist should relate himself to the larger society of which he is a part. The readings of this Section have presented only some of the diverse views held on this question by scientists. The issue, however, is a very important one, for there is every indication that scientific activity will continue far into the future, and that the findings and ideas that emerge from tomorrow's laboratories will provide fertile ground for both hope and despair, a great potential for both human power and human degradation.

In any discussion of science, a good place to begin might be the basic premise that knowledge is better than ignorance. With knowledge comes the possibility of choice, while ignorance consigns the events of our lives to accident. For whether we like it or not, we do live in a natural world where things happen according to the laws of nature, and these can be made to work either for us or against us.

Some scientists argue that they should be free to work on whatever strikes their fancy, since *any* new knowledge is worth pursuing. If such new knowledge is used by others to the detriment of the world, they say, that cannot be their concern, for knowledge is neutral—it is neither good nor evil. Only the chosen *applications* of knowledge are good or evil, and these applications are not the concern of the scientist.

Other scientists take active and partisan roles in the affairs of the world because they believe that their training and knowledge gives them special competence. Thus, for example, in the recent governmental considerations of whether or not to support the development of a supersonic transport plane, scientists were arrayed vociferously on both sides of the issue. Chemists promoted their conflicting views of the chemical effects on the upper atmosphere of a supersonic airplane: some said that the upper-atmospheric ozone content would be lowered by an amount which would be dangerous to human life, while others maintained that the effect would be too small to be significant. Another issue on which one can read conflicting "expert scientific opinion" is the effect of nuclear power plants on human health and on the environment.

When the experts disagree, what is an ordinary citizen to do? Should he simply ignore what*ever* the scientists say? Most probably, it would seem that ordinary citizens and their governmental representatives should move slowly and with caution until such time as the scientists *can* agree. For disagreements among scientists are symptoms of man's imperfect understanding of the natural world, and only further scientific study can resolve these conflicts of scientific opinion. But at the same time we can take pride in how much is already known, in how far we have already come.

Scientific knowledge accumulates by building upon itself. The scientific community operates in a most interesting way, designed to reject the false and to retain the verifiable. This system is a direct outgrowth of the veracity-curiosity features remarked upon by Kenneth Boulding in the article you have read. Wherever possible, the scientific statements made and the evidence presented by one scientist are scrutinized, repeated, tested, modified, and argued about by other scientists. Thus at the edges of science there must always be uncertainty and perhaps even a temporary confusion, but in the interior there resides a large body of knowledge which is generally believed to have been verified, and which has been found to be internally consistent through a maze of many different logical interconnections.

Is a scientist, then, just an ordinary citizen with special knowledge in a narrow region of the human enterprise, and therefore with special responsibility only for seeing that his knowledge is reliably understood by others? This question was answered affirmatively in the article by John Ziman, who would probably go on to say that a scientist also must share the same responsibilities borne by any other citizen outside his field of expertise.

Another view of a scientist's responsibilities to society holds that the scientist's discoveries always carry with them the possibility of later uses that are antisocial, and that, when such antisocial applications ensue, the responsibility falls on the scientist whether he likes it or not. While in a legalistic sense he may well argue that a contribution to knowledge, once made, is a gift to the world, and whether it is used for good or evil or both cannot be controlled by the discoverer, the moral issue remains inextricably bound up with the human concept of sin. And so it happens that when the sensitive scientist learns that his contribution to the world has its ugly side, he finds it necessary to atone for the evil even as he rejoices in the good. From him who has much to give, much is expected.

The complaint of some people who are concerned with the general welfare is that knowledge is power, and science, as a generator of knowledge, confers power on its possessor. In this context, then, the reason that governments and industries support scientific research is that it breeds power. Those individuals and groups who do not control governments or industries may believe themselves to be oppressed by this power, and for them science is one of the threatening things in the world. Such an argument has particular force when the threat consists of large sums of public money being diverted into research and development of military weapons.

An attitude toward scientific research that is not often mentioned these days is the humanistic one: scientific research is a creative enterprise in which some new perception of the world is revealed to mankind. Engaging in such creative activity, the scientist is comparable to any other creative artist. The motivation and the true reward are aesthetic first and foremost; other consequences are secondary. There are some scientists today who will argue that this is their attitude, though such scientists were more common in the past decades than now. The emphasis has shifted away from science as a form of human enjoyment to science as an aid to material progress. Just why a government will support a symphony orchestra or a creative artist as a purely aesthetic contribution to the community, while requiring a scientist to justify his work in terms of its practical significance, is not easy to see. Is it that too many scientists have been all too willing to put their own emphasis on practicality?

And what of Alan Mussett's concern that scientific endeavor, which provides valuable elements of challenge and reward in society, will soon run out of things to do? Shall we slow down the pace of scientific work so as to preserve the sense of excitement and challenge for a longer time, as Mussett suggests? Such a proposal is in direct contrast to the point of view represented by men such as Vannevaar Bush, who, after his work on the early development of computers and his scientific leadership in World War II, published a book entitled, *Science, the Endless Frontier.* But Alan Mussett now says it is not endless at all. What is the present situation in chemistry, for example? What are the possibilities of future discoveries?

New chemical compounds are being produced in chemistry laboratories each year, and there are probably five or six million compounds already known. This, of course, is a huge number, and suggests the possibility that there cannot be too many more. But surprisingly, when one tries to calculate the number of ways in which the 100-or-so chemical elements can be combined to produce compounds, one finds that five or six million is really only a scratch on the surface. While the variety of compounds yet to be discovered may not be endless, it must at least be far larger than the number now known. Furthermore, the relationship between the atomic composition of a compound and its properties is yet only imperfectly known, so that one cannot even predict with any great confidence what the properties of a new combination of elements is likely to be. Thus, even in this narrowest possible view of chemistry—the production and characterization of new compounds—we do not appear to be anywhere near the end.

Considering problems of more immediate import and larger scale than the discovery of new chemical compounds, John Platt argues for the need to direct scientific work toward heading off global crises. If this is not done, he believes, we may soon find our earth so devastated that human life will either disappear entirely or be reduced to a condition of abject poverty. Behind his concern is the assumption that events can now happen so rapidly and on such a vast scale that only planning and action well in advance of some troubles can hope to be effective. In large part this state of affairs arises because of the great and expanding size of the human population and because of the range of ways in which human actions can affect the entire earth.

The Platt argument suggests two things. One of these is that there are some matters of great complexity that are still in need of much better scientific understanding than we now possess. This degree of complexity may even be an order of magnitude greater than that of solving the problem of cancer. Yet even for a problem the magnitude of cancer, our scientific understanding is not sufficient to offer a reliable map of where to look for a cure. Our society is in the curious position of having the knowledge necessary to travel to the moon and beyond, but not the knowledge to deal with a human disease, much less with some of the problems posed by Dr. Platt. The second conclusion suggested by Dr. Platt's argument is that, in order to deal with problems of the complexity of those set forth by him, science will have more than enough to do into the far distant future. In a sense, predicting the demise of science is to underestimate the vast problems we have yet to solve.

So if John Platt is even partially right, the discussion by Dr. DuBridge in the last article in this Section, and the last in this book, is even more pertinent. The world community needs more and better scientific knowledge not only for its intellectual enjoyment, but for its very survival. And since one important resource for developing this vital knowledge is the available scientific talent, more effort should be made to insure that our human scientific talent is nurtured and encouraged, for it constitutes an important social instrument which can be used to benefit all mankind.

Just as with some other important human concerns, science is a very complex activity. It combines a creative, humanistic activity akin to the arts with intellectual excursions into the world of nature that produce discoveries not unlike those of geographic explorers. To these we must add the possibility of applications which can make the age-old dream of social justice more nearly a reality.

And so it is that we find the drama of good and evil being played with peculiar intensity in the lives and works of scientists. Only if both scientists and society at large remain constantly alert to the dual possibilities can we feel as unthreatened by the scientist as by the Arctic explorer. And only then can the scientist's soul rest as peacefully as that of the artist.

May we remain alert!

INDEX

Page numbers followed by (i) refer to illustrations. Page numbers followed by (t) refer to tables.